Identity, Contestation and Development in Northeast India

India's Northeast has long been riven by protracted armed conflicts for secession and movements for other forms of autonomy. This book shows how the conflicts in the region have gradually shifted towards inter-ethnic feuds, rendered more vicious by the ongoing multiplication of ethnicities in an already heterogeneous region. It further traces the intricate contours of the conflicts and the attempts of the dominant groups to establish their hegemonies against the consent of the smaller groups, as well as questions the efficacy of the state's interventions. The volume also engages with the recurrent demands for political autonomy, and the resultant conundrum that hobbles the region's economic and political development processes.

Lucid, topical and thorough in analysis, this book will be useful to scholars and researchers in political science, sociology, development studies and peace and conflict studies, particularly those concerned with Northeast India.

Komol Singha is Associate Professor of Economics at Sikkim University, Gangtok, India, and holds a doctoral degree in development economics. His research areas are institutional economics, social capital, political economy, rural development and development issues related to Northeast India.

M. Amarjeet Singh is Associate Professor of Sociology at the Centre for North East Studies and Policy Research, Jamia Millia Islamia, New Delhi. His research and teaching focus on conflict studies, ethnicity, identity politics and migration studies.

'The fourteen insightful chapters by scholars from anthropology, economics, history, philosophy, political science and sociology focus on a wide-ranging debate on ethnicity, identity, conflict and development of the north-eastern part of India. Students, scholars and policymakers dealing with India's Northeast will find this book useful.'

T. B. Subba, Professor and Vice Chancellor,
Sikkim University

'The book . . . engag[es] with some of the major socio-political issues in Northeast India. Ethnic identity narratives, territoriality and contestations, governance challenges, among others have been brought under one rubric for discussion. These essays would certainly generate further debates and help shape a better understanding of the region.'

Bhagat Oinam, Professor,
Jawaharlal Nehru University

'A refreshingly new take on the connection between insurgency and development by a group of young scholars beyond the otherwise clichéd script of ethnicity and identity politics . . . [A] must-read for those who think that understanding India's Northeast . . . has something to contribute to the larger social science repertoire.'

Samir Kumar Das, Professor and
Dean of Arts, University of Calcutta

Identity, Contestation and Development in Northeast India

Edited by Komol Singha
and M. Amarjeet Singh

LONDON AND NEW YORK

First published 2016
by Routledge

2 Park Square, Milton Park, Abingdon, Oxfordshire OX14 4RN
711 Third Avenue, New York, NY 10017

Routledge is an imprint of the Taylor & Francis Group, an informa business

First issued in paperback 2017

British Library Cataloguing in Publication Data
A catalogue record for this book is available from the British Library

Library of Congress Cataloging-in-Publication Data
A catalog record has been requested for this book

ISBN: 978-1-138-95419-9 (hbk)
ISBN: 978-0-8153-9585-0 (pbk)

Typeset in Galliard
by Apex CoVantage, LLC

Contents

Figures

Tables

Preface

India's Northeastern Region consists of eight states and is a 'mixed bag' of several ethnic groups. Around 200, out of 635 tribal groups of the country are found in this region. Geographically, the region covers an area of approximately 263,000 sq. km (8 per cent of the country), and its population is about 4.56 crore (3.77 per cent of the country) as per 2011 population census. Around three-fourths of the region's geographical area are covered by hilly terrain, and majority of the population (around 85 per cent) lives in villages, rural areas in poverty, in the midst of abundant resources. The region is connected with the mainland India by a small strip of land, around 20-km-wide chicken's neck corridor at Siliguri in West Bengal. Physical infrastructure of the region is very poor compared to national level.

Despite its rich natural and human resources, the region is mired by a series of intractable conflicts. In the recent past, it has slowly shifted towards the inter-ethnic feuds and rendered more viciously by the ongoing amoebic multiplication of ethnicities in an already prodigiously heterogeneous region. Strident ethno-nationalistic assertions over the land and resources and articulation of grievances in terms of the 'others' have given rise to contestations over the same space shared by multiple ethnic groups. Tacit assertions and explicit demands for hybrid identity-based exclusive homelands are accentuating the latent sociological fissures into apparent fault lines and often lead to ethnic cleansing and extermination of the smaller groups or politically weaker sections in the society. The intricate contours of the conflicts are getting convoluted with the attempts of the dominant groups to establish their hegemonies, and that are being challenged by the emerging subalterns sharing the same space. Propensity of the stakeholders to use violence as a political resource in pursuit of their objectives has fostered the ecology of violence resistant to restraint. State's interventions to contain the non-state violence with rewards and incentives along the lines of ethnicities have compounded the situation further, accelerated the growth of newer identities or hybrid identities, and it often leads to recurrence of demand for autonomy within the region. The resultant conundrum hobbles the region's economic and political development processes in the long run.

Having understood this fact, many scholars working on the issue were invited to share their scholarly works, and finally fourteen essays have been chosen for this volume. The present volume entitled *Identity, Contestation and Development in Northeast India* is outcome of it. Though the eight states of the region have different issues and challenges, these chapters made a modest attempt to understand the issue of identity, contestation and overall development of the region. How has the ethnic politics complicated development initiatives and 'take-it-easy' approach of the government compounded conflict situations in the region, are well debated in this volume.

Definitely, this volume will serve as a landmark in the region's literature on ethnic identity formation, politics of state formation and development policy formulation. It will also be very useful for the scholars, academia and policymakers for further research of the region's complex development issues. Therefore, we would like to express our heartfelt gratitude to all the contributors for their valuable contribution in this volume. It would not have been possible without their support. We would also like to thank R.N. Ravi, Former Special Director, Intelligence Bureau of India, and Prof S. Japhet, National Law School of India University, Bengaluru, for their encouragement in inception of this volume. Last, but not the least, our special thanks goes to Routledge for bringing out this volume on time.

Contributors

P.R. Bhattacharjee is retired professor of economics at the Assam University, Silchar (Assam). Prior to Assam University, professor Bhattacharjee had served in Tripura University. His works were mainly centred on the issues of growth and development of Northeast India.

Rakhee Bhattacharya holds a doctoral degree from Gauhati University and is currently associate professor at Jawaharlal Nehru University, New Delhi. She is working on various economic challenges of contemporary India. She has published widely on the issues of insurgent movements of Northeast India, inclusion and voices of the underprivileged and marginalised sections.

Sarit Kumar Chaudhuri is director of Indira Gandhi Rashtriya Manav Sangrahalaya, Bhopal (Madhya Pradesh). He is the recipient of Young Scientist award from Indian Science Congress Association, and was post-Doctoral Fellow for two years at SOAS (School of Oriental and African Studies), University of London. He has also worked at Anthropological Survey of India.

Tajen Dabi is assistant professor of history at Rajiv Gandhi University, Itanagar (Arunachal Pradesh). Dabi's research and teaching focus on medieval and modern Indian history, history and politics of Northeast India and historical linguistics.

Mohan Debbarma is head and associate professor at the Department of Philosophy, Tripura University, Agartala (Tripura). Debbarma's works centre on the issues of inequality, conflict and indigenous communities, especially the Borok (Tripuri) of Tripura.

Uddipana Goswami is editor of *Northeast Review*, an online journal of arts and literature, Guwahati (Assam). She has published extensively on the conflict and development issues of Northeast India. Her published works include a poetry collection, *We Called the River Red: Poetry from a Violent Homeland* (2010), and an edited volume, *Indira Goswami: Passion and the Pain* (2012).

Kilangla B. Jamir holds a doctoral degree in economics from North-Eastern Hill University and is currently professor of economics at the Nagaland University, Lumami (Nagaland). She has been working on the development issues in Naga communities of Nagaland, and published extensively on the issue.

Lalrintluanga is professor and head of the Department of Public Administration, Mizoram University, Aizwal (Mizoram). Lalrintluanga's research works focus on district administration, administrative theory, development administration, comparative public administration and women's empowerment.

Purusottam Nayak holds a doctoral degree from IIT Kharagpur and is currently professor at the Department of Economics, North Eastern Hill University, Shillong. Nayak's research and teaching focus on development economics, human development and inequality of communities of Northeast India.

J.J. Roy Burman, an anthropologist, is currently professor at the Centre for Study of Social Exclusion and Inclusive Policies, Tata Institute of Social Sciences, Mumbai (Maharashtra). He also carries out research programmes on the problem of rural areas and weaker sections. He is the first person in India to have completed PhD on the sociopolitical dimensions of sacred groves.

L. Muhindro Singh is associate professor at the Human Rights Studies Centre, S. Kula Women's College, Nambol (Manipur). He was senior fellow, Indian Council of Social Science Research, New Delhi. He has been writing on the issues of identity, politics, conflict and developments of Northeast India.

Hoineilhing Sitlhou holds a doctoral degree in sociology from Jawaharlal Nehru University and is currently assistant professor of Sociology at the University of Hyderabad (Telangana). She has done extensive research work on the issue of Thadou-Kukis of Manipur. Among other academic achievements, she has presented research works in several national and international conferences.

Introduction

India's Northeast, the confederation of eight states (i.e. Assam, Arunachal Pradesh, Manipur, Meghalaya, Mizoram, Nagaland, Sikkim and Tripura), is home to around 200 indigenous communities (Cline, 2006). Geographically, the region covers an area of approximately 263,000 sq. km (8 per cent of the country), and its population is about 4.56 crores (3.77 per cent of the country) as per 2011 population census. Around three-fourths of the region's geographical area are covered by hilly terrain. The region shares long land borders with as many as five countries, viz. Bangladesh, Bhutan, China (Autonomous Tibet Region), Myanmar and Nepal, and is situated at the tri-junction of South, East and South-east Asia. It is also connected to the 'mainland' India by a narrow 20-km-wide chicken's neck corridor at Siliguri town of West Bengal.

Ethnic communities living in the Northeastern Region (NER) are seemingly homogenous to laymen from outside but are in reality heterogeneous in many respects – socially, linguistically, culturally, religiously and politically. The state of physical infrastructure that includes road and communication, water, power, health, education and so on is found to be very weak compared to other parts of the country. Inter- and intra-state connectivity of the region is a challenge. This remoteness and weak connectivity could be one of the factors of perceived sense of isolation from the 'mainland' India. Its overall socio-economic structure is also characterised by inadequate transport and communication facilities, limited industrial activities, high unemployment, limited educational facilities and low per-capita income. These constitute a significant factor in studying conflict and for overall economic backwardness in this region. At the same time, New Delhi's 'take it easy approach' has convoluted the situation of conflict and ethnic relation of the region. Now the region has become an epicentre of ethnic conflict, identity conflict, armed conflict and violence in the country.

Soon after India's independence, the Nagas demanded an independent homeland, but it was denied to them. An attempt was made in mid-1947, when the Governor of Assam and the Naga National Council (NNC), a pro-independence group, agreed to set up an interim administrative arrangement in the Naga Hills. However, the proposal could not materialise as the agreement became contentious. The most controversial clause was 'The Governor of Assam as the Agent of the Government of Indian Union will have a special responsibility for a period of ten years to ensure that due observance of this agreement; at the end of this period, the Naga National Council will be asked whether they require the above agreement to be extended for a further period, or a new agreement regarding the future of the Naga people would be arrived at.' The Nagas insisted upon complete autonomy and independence once the interim period of 10 years was over, while India insisted that it only means the right to suggest administrative changes under its constitutional framework. Thereafter, India offered limited autonomy under the Sixth Schedule of its constitution, which was rejected by the Nagas. The first general election held in 1951 was boycotted in the Naga Hills, marking the beginning of the first armed struggle against India. Later on, a few pro-India Naga leaders proposed for setting up of a single administrative unit with the Indian union comprising Tuensang Division and the Naga Hills. It was readily accepted by India (hereafter, the centre/New Delhi). In 1959, the pro-India Naga leaders put up another proposal for the elevation of Tuensang Division and the Naga Hills into a state of India, and eventually, the area became the state of Nagaland in 1963. It was, however, opposed by the pro-independent NNC and hence started an armed conflict against India. Further, in 1975, an NNC faction and New Delhi signed an agreement in which the former agreed to abide by the Constitution of India and surrender arms. Unfortunately, it was rejected by another NNC faction for what they called 'an agreement to sell the Naga nation'. As a result, in 1980, the breakaway group founded the National Socialist Council of Nagaland (NSCN). It later split into two – NSCN-IM and NSCN-K. Although the NSCN-IM has been able to prove itself as the most influential one, the prospect for a long-lasting solution is not forthcoming because the demand for bringing the Nagas together into a single politico-administrative unit will not go unchallenged from other ethnic groups.

Like Nagaland, Manipur has also been affected severely by protracted conflict since the 1960s, which is primarily a political in origin over the questions of autonomy, independence and ethnicity. Historically, Manipur was one of the oldest independent kingdoms in South-east Asia, having its own civilisation, traditions and cultural heritage (Verghese, 2012). It was ruled in succession by different kings since 1445 AA until 1949, and it was well recorded in the Royal Chronicle *the Cheitharol Kumbaba* (Tensuba, 1993). Manipur came under the British rule in 1891 and annexed to the Indian Union as part 'C' state[1] on 15 October 1949 (Sharma, 2011; Suan, 2009). Thereafter, it was directly ruled from New Delhi, and the bureaucrats of the centre were not trusted by the local people (Rammohan, 2002). Thus, a movement resisting the merger (annexation, as termed by Meiteis) started and subsequently culminated into an armed conflict. Finally, Manipur became a state of Indian union in 1972. But, it failed to end the armed conflict. Manipur is now one of the most restrictive places where the extraordinary legislations like the Armed Forces (Special Powers) Act (AFSPA) of 1958 and the Foreigners (Protected Areas) Order of 1958 are in operation.[2] Successive governments have not been able to tackle the problem leading to the erosion of democracy, human rights and rule of law, which resulted in a series of flashpoints. Irom Sharmila (Iron Lady) has been fasting since 2001, demanding for the repeal of the AFSPA. She is kept alive by forced feeding at a public-funded hospital in Imphal, Manipur's capital city. The alleged rape and murder of another lady, Thangjam Manorama, in 2004 by the security forces sparked a wave of protest including a nude protest by a dozen women in Imphal.

Ironically, the political aspirations of Manipur's main ethnic groups – the Meiteis, the Nagas and the Kukis – are at loggerheads with one another. On one side, the Meiteis, the dominant community who live in the plain want an independent homeland. On the other side, the Kukis and the Nagas who live in the hills surrounding the plain want bifurcation of the hill portion from the Meiteis[3] and further division of the hills into two – one each for the Nagas and the Kukis. This is with the objective of enjoying political control over the territories that they dominate. The Nagas call their homeland as Nagalim, while the Kukis call their homeland as Kukiland. As a result of this, a series of ethnic conflict had happened in the 1990s due to overlapping homeland demands by the Nagas and the Kukis. Even ethnicity-based political party has emerged, for example the Naga People's Front.

By contrast, at the time of the Partition of India (1947), a large number of refugees (the Hindu Bengalis) fled towards Tripura, which became a cause for hostilities between the older inhabitants known as the Boroks (Tripuris) and the refugees (Bengalis). Within a few years, the Boroks, erstwhile majority community, were reduced to minority, outnumbered by the Bengalis. Being the single largest group, at present, the Bengalis are elected from at least two-thirds of the constituencies in the legislative assembly; this number is sufficient enough to control the state government. Bengali and English became the official languages in 1964, but 15 years later, in 1979, Kokborok, the language of the Boroks (Tripuris), became an additional official language (currently, Bengali and Kokborok are the official languages) in Tripura. The Boroks blamed Bengali hegemony for their backwardness and accused local government (control by Bengalis) of imposing the learning of Bengali upon them.[4] As a result of this, three decades later a deadly riot took place between Tripuris and Bengalis.

Like many other conflicts around the world, land and resources are the central issues in Tripura too. The Boroks have been practicing jhum (shifting) cultivation for ages. A significant number of families still continue to do so, and they are known as Jhumias. In order to discourage jhum cultivation, several steps were taken up by the government. The last king, Bir Bikram Kishore Manikya, kept a large tract of land in Khowai, called the Kalyanpur Reserve, for the settlement of the Jhumias. After independence, this land was allocated to the refugees. With the passage of time, the refugees or Bengalis became policymakers, politicians, ministers, moneylenders and rich farmers. The Boroks mortgaged land to obtain loan, but when they failed to pay interests in time the moneylenders took the advantage to acquire the mortgaged land. As a result, they were gradually alienated from their ancestral land and driven away to far-flung areas.

Like in Tripura, large-scale immigration into Assam started when the British introduced modern administration and started tea plantation, coal mining and oil exploration. Consequently, the identity politics started when Assam was placed under Bengal province since 1826 (continued to be ruled from Bengal till 1873). It was only in 1874 that the whole of erstwhile Assam was separated from Bengal and made a Chief Commissioner's province by incorporating a major portion of Bengali-speaking areas of Cachar, Sylhet and Goalpara.[5] Besides, instead of Assamese, Bengali became the language of administration and education before Assamese got its dues after several decades. These developments significantly contributed to hostilities between the Assamese and the Bengalis (GoA, 2012).

As of the natural resources, Assam is one of the single largest producers of natural gas, oil and tea in the country, but the alleged over-exploitation of these resources is a cause of concern. Local people alleged that Assam served mainly as the supplier of raw materials and the market for finished goods. Assam protested against New Delhi's decision to set up an oil refinery at Barauni town in Bihar to refine crude oil extracted from Assam. Once the refinery was built at Barauni in 1962, an oil pipeline connecting Guwahati and Barauni was hurriedly constructed for the transportation of crude oil. After a prolonged agitation a small oil refinery was set in Assam. This shows that centre listens only when Assam revolted against for their grievances (Hrishikeshan, 2002). Many more people also alleged that the royalty from crude oil and natural gas production was negligible, and 'Assam has to request, cajole or threaten the centre to pay or hike the rate of oil royalties due to the State' (Hussain, 2013). Again, most of the tea gardens were owned by the outsiders, and their corporate offices were located in the far away cities outside Assam. It was alleged that the profit was not utilised for the development of the state (Gogoi, 2007). In 1976, a committee of the Assam government noted:

> A comparative study between the gardens under four different associations [the Assam Branch of the Indian Tea Association, the Tea Association of India, the Bharatiya Chah Parishad, and the Assam Tea Planters Association Brahmaputra Valley] reveals that, so far as the representation of local people in the managerial cadre (including Assistant Managers) is concerned, the position is the worst in respect of the tea gardens under the Tea Association of India, where 82 per cent of posts in the managerial cadre have been held by persons with birthplaces outside Assam . . . One striking point to be noted is that, after change-over to Indian management, almost all the posts in the managerial cadre (including Assistant Managers) were filled up by persons from outside the State of Assam without any open advertisement or without notifying the employment exchange. (GoA, 1976)

When the Chinese forces entered Assam in November 1962, during the Indo-China war, the Indian forces deserted the town of Tezpur. At that time, Prime Minister Jawaharlal Nehru had said in a radio broadcast, 'My heart goes out to the people of Assam, but I cannot do anything' (Saikia, 2004). The people were 'shocked and devastated' by this statement (ibid.: 166), and Nehru's message was interpreted as 'a parting message to the Assamese people' (Gogoi, 2007). The feelings of neglect and discrimination are reinforced by poor economic growth during the post-independence period.

Furthermore, during the Bangladesh liberation war[6] of 1971, millions of refugees fled to India, mainly in Assam. Fearing the atrocities in Bangladesh, many of them did not go back after the war and settled in Assam, Meghalaya and Tripura. Even thereafter, immigration driven purely by economic consideration has taken place. Bangladesh is one of the most densely populated (964 persons per sq. km in 2011) and poorest countries in the world. In addition, the country is regularly affected by natural disasters such as flash flood, riverbank erosion and landslide displacing a large number of people. This led to perennial influx of a large number of migrants into Assam and other neighbouring states. Yet, no one really knows the exact population of immigrants and refugees after the partition or the creation of Bangladesh. Neither India nor Bangladesh maintains any reliable records. However, the popular perception is that large-scale immigration is a demographic threat to the survival of small ethnic groups. This is further reinforced by the fact that the immigrants from other parts of India had control over the trade and commerce, agriculture, household labour and manual labourers in Assam. Understandably, the illegal immigration is also considered by the authorities, media and vigilant pressure groups as a 'major challenge', 'main problem', 'cultural threat', 'security threat', 'silent demographic invasion' and 'conspiracy' (MHA, 2013; Kumar, 2006). The notion of a myth of large-scale immigration still persists and is reinforced by vigilant movement against immigration. Immigration is said to have caused higher population growth, with Assam showing a decadal population growth rate higher than the all-India average during the major part of the twentieth century. The higher growth of Muslim population in Assam in the recent past, higher than the state average, has been extensively debated upon, due to immigration (e.g. Dutta, 2012), while the Chief Minister of Assam, Tarun Gogoi, in 2012[7] says that it is due to higher illiteracy among the Muslims. There are also allegations of the enrolment of illegal immigrants in the voters' lists.

In 1960, the Assam Official Language Act of 1960 was passed. It authorises the use of Assamese for all or any of official purpose of the state, while Bengali in the Cachar district and English in the autonomous districts. Despite this measure, in the State, non-Assamese people perceived it as the Assamese hegemony. Further, in 1972, the Gauhati University, the oldest university in the region, proposed to make Assamese as the language of instruction in educational institutions under its jurisdiction. Likewise, in the late nineteenth and early twentieth centuries, the Bengali Hindus being politically the most significant group had used the government to consolidate their positions in the educational system. In the 1930s and 1940s, when electoral politics were introduced, the Bengali Muslims had control over the government and attempted to use their positions to facilitate immigration of Muslims into Assam. After independence, Assamese utilised their dominance in the government to consolidate their language in the economy and social system (Weiner, 1989).

Right after the language issue was partly settled, the focus swiftly turned towards immigration and electoral politics in Assam. The death of Hiralal Patwari (Member of Parliament from Mangaldai parliamentary constituency) in 1979 required the holding of by-election. In the process, it was noticed that the number of voters in this constituency had risen abnormally. At this juncture, the main student group, All Assam Students' Union (AASU), demanded a rescheduling of election and the name of foreign nationals be deleted from the electoral rolls. This marked the beginning of a mass movement known as the Assam Movement. The movement demanded recounting the citizenship of those living in Assam on the basis of the National Register of Citizens which was prepared during the Indian Census of 1951. Spearheaded by AASU, the movement asked the centre to identify all the 'foreigners', remove their names from the electoral registrar and deport them back to their home country. The issue slowly concluded with the signing of an agreement, *the Assam Accord*, between the centre and the leaders of the movement in 1985, which promised to take appropriate actions to identify and deport all the 'foreigners' who came to Assam after March 1971 and to disenfranchise those who came between January 1966 and March 1971. It also promised constitutional, legislative and administrative safeguards to protect and promote the culture, social, linguistic identity and heritage of the 'Assamese people'. But, these promises have thus far remained unfulfilled.

The movement coincided with the formation of United Liberation Front of Asom (ULFA) in 1979 with the goal of complete independence for Assam through an armed struggle. It insisted that Assam was never a part of India and hence the relationship between the two was 'colonial' (see Mahanta, 2012). Its founders hoped that the Assam Movement was not enough because New Delhi would not succumb to mere strike and shouting. From a more militant stream of Assam Movement it broke away from the moderate forces that were associated with it (Das, 2007). Down the line, it has softened the hard-line stance towards Bangladeshi immigrants claiming that they are an indispensable part of Assam (ibid.: 14). Instead, it started viewing the Hindi-speaking people from other part of country as more disturbing and had even warned them to 'go away'. They claimed that 'once freed of Indian bondage they would be able to solve the problem [immigration] with ease and dispatch' (Gohain, 2007). After the Assam Movement, its young leaders set up a political party called Asom Gana Parishad (AGP) in 1985, while those who were against the Assam Accord formed another political party called United Minorities Front (UMF). The former wanted the implementation of the said accord, whereas the latter was against the accord. The AGP won the legislative elections of the state held in 1985 and 1996 but failed to solve the problem that they had been fighting for. Although the UMF had been relegated to the margins, a new political party – All India Democratic United Front (AIDUF) – has now become an important political player, and they both share the same ideology.

With the aim of checking illegal immigration into Assam, the country's parliament enacted a special legislation as well as a slew of border security measures. In 1983, the parliament passed the Illegal Migrants (Determination by Tribunals) Act to detect and deport non-citizens, officially referred to as the 'foreigners', whereas the Foreigner Act of 1946 was enforced in the rest of the country. In addition, the policing along the border of India and Bangladesh was strengthened with the construction of a border fence, roads and floodlighting. However, India is unable to enact suitable legislation acceptable to all sections of the population. For instance, the promulgation of the Illegal Migrants (Determination by Tribunals) Act had caused division among the local people. The Muslim groups welcome this law, while others (locals) were against it. When it was struck down by the Supreme Court in 2005, the Muslim groups alleged that the Muslims would be harassed by the police in the name of detection of 'foreigners'. By contrast, the other groups who were against this law welcomed the ruling. Thus, the public opinion is polarised on the issue of immigration, and ethnic polarisation and conflict continue. In reality, the Assam Accord was not free from criticism. Non-Assamese-speaking people objected to one of the provisions that promised to protect and promote the culture, social, linguistic identity and heritage of the 'Assamese people'. This became a rallying point in the relationships between majority Assamese and other smaller ethnic groups. Some of the smaller ethnic groups who have somehow culturally assimilated into the dominant Assamese culture have started reasserting their separate cultural and ethnic identities. For instance, the Bodos demanded a separate state to be carved out of Assam. Their movement ultimately led to the granting of limited political and economic autonomy under the Sixth Schedule of the constitution in four contiguous districts of Kokrajhar, Baska, Udalguri and Chirang known as Bodoland Territorial Areas District (BTAD) in 2003. But this settlement opened up series of violence, rather than taking care of the non-Bodos.

Like other non-Assamese-speaking people, the Mizos protested against the declaration of Assamese as the official language in 1960. The anger was further fuelled by inadequate response to famine (known as *Mautam*) which occurred in Assam's Lushai (Mizo) hill district. It was caused by the destruction of standing crops by rodents (the rodents grew exponentially due to bamboo flowering). Since the government's relief works were inadequate, several voluntary organisations including Mizo National Famine Front carried out relief work. It subsequently dropped the word 'Famine' and transformed into an armed group, known as Mizo National Front (MNF) seeking to achieve an independent homeland. Finally, centre signed an agreement (Mizo Accord in 1986) with MNF to recognise Mizoram as a separate state and the latter transformed into a political party. With this, Mizoram is now considered as one of the most peaceful states in the country. But, in reality, it was settled by oppressive manner after the Indian Air Force bombed Mizoram (Ngaihte, 2013). Besides, thousands of Brus and Chakmas have been driven away to Tripura in 1997, and the minority communities have been crushed by militant groups, in part by Mizo Youth Organizations acting as an extra-legal police force (Lacina, 2009).

When we analyse in entirety, to contain conflict in NER, centre relied on three-fold strategies – (1) a strong security approach, (2) expansion of the avenues for political dialogues and (3) allocation of more funds. Since the problem of the region is seen primarily from a law and order approach, the use of armed force is legitimised. The armed forces used by the centre are also not accountable to the regional governments for the abuse of power. The political strategy concentrated on the creation of new states and autonomous regions. When the internal boundary of India was reorganised in the mid-1950s it was believed that Assam should not be divided. But the pressures from various ethnic groups compelled Assam to bifurcate into different parts. In 1963, the Nagas got Nagaland carved out of Assam. Assam was further divided in 1972 to create Meghalaya, Mizoram and Arunachal Pradesh. Finally, huge funds are being pumped into this region in the name of accelerating development. These states are categorised as Special Category States, into which centre funds up to 90 per cent of their capital budget requirements. There is also a separate ministry called Ministry of Development of North Eastern Region (DoNER), which deals with the planning, execution and monitoring of development projects of the region. Despite these efforts, the region is mired by a series of intractable conflicts and consequently ransoms economy.

Ethnic identity and conflict

What is peculiar to this region in the recent past is – the salience of ethnic identity as a means of political mobilisation and conflict. Hence, this section gives an introductory overview of ethnicity and identity. The meaning of *identity* varies from one scholar to another, although in broader sense, it is 'people's concepts of who they are, of what sort of people they are, and how they relate to others' (Hogg and Abrams, 1988). Social identities are first steps to political identities. Religion, national and ethnic identities generate powerful emotions and hence have political importance (Berezin, 1999). Since identity can be a source of pride and joy, many of the conflicts are sustained through the illusion of a unique and choice-less identity (Sen, 2006). Nevertheless, identity is an important element in the development of nationalism and conflict. According to Esman (1975), the proportion of conflict and cooperation depends on the *relative resources* at the disposition of each group. These resources are demographic (relative numbers); organisational (degree of mobilisation and capacity to put resources to political uses); economic (control of finance, means of production or trade channels); technological (possession of modern skills); locational (control of natural resources and strategic territory); political (control or influence over the instrumentalities of the state) and ideological (the normative basis for group objectives). In addition to these objective determinants of power, the quality of inter-communal relations depends on the *congruity* or *disparity* in goals between those who control the state apparatus and the leaders of the constituent groups. If the goals are same, the outcome is likely to be consensual. If the goals are incompatible, the consequences will have tension and conflict, and the outcome will be determined by the relative resources controlled by the parties. This introduces to a third determining factor – *the institutions for conflict management*, the conventions, rules, procedures and structures of conflict. Without such institutions, there can be no predictability in inter-group relations and no framework for channelling group demands or for regulating outcomes (ibid.: 392).

The ethnic group formation is thought to be a process that involves several steps, as indicated by the aforementioned discussion. The first step takes place within the ethnic group itself for control over its material and symbolic resources, which in turn involves defining the group's boundaries and its rules for inclusion and exclusion. The second step takes place between ethnic groups as a competition for rights, privileges and resources. The third step takes place between the state [nation state] and the groups that dominate it on the one hand, and the populations that inhabit its territory on the other (Brass, 1991). Ethnicity is central to individual and group identity. It can provide an important thread of continuity with the past and is often kept alive through the practice of cultural tradition. At the same time, it is fluid and adaptable to changing circumstances (Giddens, 2006). But, with rapid modernisation, technological advancement and increased mobility, it is now possible to choose one's ethnic identification in a self-conscious way (Hutnik, 1991). Following polarisation, thc cthnic conflict constitutes a kind of implicit bargaining, even if the participants do not think so (Banton, 1998). The clustering of factors that give rise to ethnic political mobilisation is complex, may vary from case to case. However, in developing countries like India there are several commonalities. As of the causes of conflict, Ganguly (2009) identifies four sets of causal conditions which have usually combined in different ways to produce ethnic conflict in India. They are (1) the fear that assimilation could lead to cultural dilution, and the unfulfilled national aspirations; (2) the process of modernisation has sharpened their sociopolitical awareness and increased their capacity to mobilise for collective action; (3) unequal development, poverty, exploitation, lack of opportunity and threats to the existing group privileges and (4) political factors such as endemic bad governance, the growth of anti-secular forces, institutional decay and vote-bank politics on the part of the unscrupulous political parties and politicians. Similarly, Olzak and Nagel (1986) provided four propositions for ethnic mobilisation: (1) urbanisation increases contact and competition between ethnic population; (2) expansion of industrial and services sectors of the economy increases completion among ethnic groups for jobs; (3) development of peripheral regions or the discovery of resources in a periphery occupied by an ethnic population and (4) processes of state building (including those following colonial independence) that implement policies targeting specific ethnic population increase the likelihood of ethnic collective action.

Land is an important element of ethnic identity. In many societies, land and identity are inextricably linked. The history, culture and ancestors of communities are tied up with land (Dale and McLaughlin, 1999). Due to its economic, social and emotional importance, land is also an important source of power. Perceived threats to security, livelihoods or identity can mobilise communities to engage in conflict (United Nations, 2012). In reality, land alone is not the sole cause of conflict; it is only a contributing factor. Land is also often used interchangeably with territory that has been identified and claimed by a person or people.

Territory is a crucial component of ethnicity because the ethnic group is usually attached to a specific territory (Penrose, 2002). Thus, the concept of territory encompasses not just geographic space but also mechanisms of authority and rights. Thus, territoriality means 'the attempt to affect, influence, or control actions and interactions (of people, things, and relationships) by asserting and attempting to enforce control over a geographic area' (Sack, 1983). In other words, the control of space is an extremely important component of power relations. When people create territories, they create boundaries that both unite and divide space along with everything that it contains. By combining some people and certain resources and separating them from other people and other resources, the creation of territories gives physical substance and symbolic meaning to notions of 'us' and 'them' and 'ours' and 'theirs' (Penrose, 2002). In this manner, the territory becomes a homeland precisely because it is to be commonly defended, because all group members share similar obligations for its protection and because it defines who 'we' are (Goemans, 2006). A homeland may be external to the country of residence of the group, or only a portion of a country or may span the territory of more than a country. But, the homeland is a 'perception, susceptible to change over time' (Toft, 2005). Thus, a homeland is a special category of territory, not an object that can be exchanged, but an indivisible attribute of group identity. This feature explains why ethnic groups rationally view the right to control their homeland as a survival issue, regardless of a territory's objective value in terms of natural or man-made resources. Homeland control ensures that a group's language can be spoken, its culture expressed and its faith practiced (ibid.: 1–45). Ethnic groups seek to rule territory in which they are geographically concentrated, especially if that region is a historic homeland. They will show little interest in controlling territory when they are either widely dispersed or are concentrated only in cities. For them, territory is often a defining attribute of collective identity, inseparable from its past and vital to its continued existence as a distinct group (Smith, 1986). Territorial attachments and people's willingness to fight for territory appear to have much less to do with the material value of land, and much more to do with symbolic role it plays in constituting people's identities and providing a sense of security and belonging (Walter, 2006).

Identity can also be the consequence of policies and acts of powerful agents, states and dominant groups who define groups by assigning labels and treating them differently over generations (Gurr, 2002). Ethnic identities are enduring social constructions that matter to the people who share them. How much they matter depends on people's social and political circumstances. Ethnic identity leads to political action when ethnicity has collective consequences for a group in its relations with other groups and states. When ethnic identity is highly salient, it is likely to be the basis for mobilisation and political action (ibid.: 6). In divided societies, ethnic affiliations are powerful, permeate, passionate and pervasive (Horowitz, 1985). According to Gurr (2002), the salience of ethnic identity at any point of time is mainly due to three factors: (1) the extent to which they differ culturally from other communal groups with whom they interact; (2) the extent to which they are advantaged or disadvantaged relative to other groups and (3) the intensity of their past and ongoing conflicts with rival groups and the state. Ethnic groups can become more or less inclusive. Some small ethnic groups merge with or absorb to other, or are absorbed by them, producing larger, composite groups. Larger groups, on the other hand, may divide into their constituent parts, or a portion of such a group may leave it to form a new, smaller group. Group boundaries thus grow wider or become narrower by processes of assimilation or differentiation. New groups are born, though old groups do not always die when this occurs (Horowitz, 1985).

A conflict is ethnic if the contenders identify themselves using ethnic terms (Stanvenhagen, 1994), and there is evidence to suggest that the 'control of the state is a central ethnic conflict objective' (Horowitz, 1985). While conceptualising ethnic conflict, different approaches have been debated upon. Firstly, the instrumentalists view ethnic conflict being driven by either the relationship between economic wants – greed and grievance – or the active manipulation of ethnic identities by political leaders for their political gain. Secondly, the constructivists attribute it as a product of historical processes over time that results in divergent ethnic identities and hostility between them. Thirdly, the primordialist theory stems from ancient hatreds between ethnic groups and that frustration comes with differences in natural ties that derive from religious, racial or regional connections (Weir, 2012). According to Baqai (2004), the ethnicity syndromes prevalent in South Asia are mainly primordial in nature. But, Dudková (2013) rejected the primordialist theory as it stresses the uniqueness and overriding importance of ethnic identity. According to him, biological characteristics, religion, language and other such characters are powerful factors that may produce ethnic conflicts in a multicultural society. According to Weiner (1989), weak modern political institutions and their inability to deal with the local religious pressures, linguistic differences and unequal power and resource sharing led to ethnic conflict in the region.

Dwelling on to the ethnic identity and conflict in India's NER, Dudková (2013) opined that granting autonomy and statehood for containing conflict, without employing consociational principle on the new states, led to the emergence of new cleavages and ethnic conflict within the newly formed states. If autonomy is given in the name of dominant ethnic group, it will make further cleavages among ethnic communities in the region (Haokip, 2012). Through their political power, the majority of the ethnic groups constantly resort to ethnic cleansing to demonstrate their control and authority (Deori, 2013). In this regard, Singh (2013) proposed a way to satisfy the aspirations of different ethnic minorities without disturbing the existing state boundaries, the 'cultural autonomy', which includes religion, culture, language, social practices, customary law and some other welfare measures. The negative effect of ethnicity can only be attenuated by the consistent application of democratic principles, norms and values through socio-economic development programmes, and the role of state is very crucial (Baqai, 2004). According to Wright (1999), in a multicultural society, a more constructive solution between majority and minority may be achieved if attention is focused upon the nature of pluralistic democracy, and the state honestly intends to achieve it, in accordance with the Charter of the United Nations, rather than pursuing the recognition of some elite community or groups.

Conflict and development

All forms of social and political movement affecting NER at present are primarily ethnically driven over self-determination, secession or political dominance. At the same time, when one looks very closely, the clustering factors which caused these movements differ from one another. The root cause of some of them can be traced back to the alleged forcible integration into Indian fabric (e.g. Manipur, Naga Hills), while others can be traced back to the question about large-scale immigration and refugees and lack of development (e.g. Assam, Tripura). In response to these movements, Assam was reorganised thrice to facilitate the formation of three new states (Meghalaya, Nagaland and Mizoram), and it also enacted preferential policies for the region. However, such policies have not been able to address many of the long-standing and deep-rooted issues and challenges, and hence there is ample evidence of the effect of conflict on overall development and well-being of the people.

Yet, studying the cost of conflict is not easily defined and hence is a herculean task. Therefore, the cost can be seen in terms of loss of human lives, displacement of people, destruction of properties and other social and economic costs. Any political or social movements and their associated armed groups, as in NER, require a large amount of money (also people, ideology and motivation) to sustain them over a particular period of time. Where does the money come from? Obviously they levy 'taxes' from local business community, public and private companies, politicians and salaried people. The best-kept secret practiced in NER is that local retailers fraudulently minimised losses caused by such taxes and other disruptive activities by creating artificial scarcities of essential commodities and selling at higher prices later.

Apart from the armed conflict between the state and non-state actors, the contesting claims over land and territory by different ethnic groups, backed by armed groups, have led to violent conflicts. These conflicts have claimed many innocent lives and displaced many more, thereby disturbing their livelihood for years in the region. The nature of conflict which was previously confined between the state and non-state armed groups has extended into the inter-group hostilities in the recent past in NER. As a result, several bloody ethnic riots have taken place. For instances, we recalled the bloody riots between Muslim immigrants and Bodos in Assam, Kukis and Nagas in Manipur, Paites and Kukis in Manipur, Khasis and Bengalis in Meghalaya, Brus and Lushais in Mizoram, Bengalis and Tripuris in Tripura and so on. This protracted conflicts and violence led the overall development process of the region to ransom. Development works suffer due to frequent strikes, insecurity, violence and levying of taxes. Armed groups interfere in the award of major contracts by government agencies and thereafter at the time of implementation. Rail infrastructure, oil pipelines and installations, roads and bridges were also targeted. Funds meant for development projects were siphoned off to support armed groups by corrupt officials. Officials of crucial government agencies colluded with them due to fear of intimidation, and in turn, this inseparable relationship promoted corruption. The presence of armed state and non-state actors restricts people's daily activities, and this in turn affects their work culture and productivity. In such state of affairs, private investments are being discouraged. The presence of such conflicts has been identified as one of the reasons for 'very slow' progress of railway modernisation works in the region, particularly in Assam (Government of India, 2005). As a result, the government's attention has been diverted away from development activities to enforcement of law and order situation in the region. Thus, there is tardy progress of development activities in areas affected by violent conflicts. It is noticed that relatively peaceful areas of the region have grown at the faster pace (e.g. Mizoram, Sikkim) in various areas such as per capita income, per capita consumption of electricity, literacy and health care in comparison with areas affected by various forms of conflict (Singh, 2011).

In a nutshell, despite its rich natural and human resources, the region continues to be affected by protracted conflicts. But, it has slowly shifted towards internal feuds, rendered more viciously by the ongoing amoebic multiplication of ethnicities in an already prodigiously heterogeneous region. Strident ethnonationalistic assertions over land and resources, and articulation of grievances in terms of the 'others' have given rise to contestations over the same space shared by multiple ethnic communities. Tacit assertions and explicit demands for hybrid identity-based exclusive homelands are accentuating the latent sociological fissures into apparent fault lines, and it often leads to ethnic cleansing and extermination of smaller communities or politically weaker sections in the society. The intricate contours of the conflicts are getting convoluted with the attempts of the dominant communities to establish their hegemonies, and when that are being challenged by the existing and emerging subalterns sharing the same space. Propensity of the stakeholders to use violence as a political resource in pursuit of their objectives has fostered the ecology of violence resistant to restraint. State's interventions to contain the non-state violence with rewards and incentives along the lines of ethnicities have compounded the situation further, accelerated the growth of newer identities, and it often leads to recurrence of demand for autonomy within the region. The resultant conundrum hobbles the region's economic and political development processes in the long run.[8]

Notes

1 Part C states include both the former Chief Commissioners' Provinces and some princely states. Each was governed by a Chief Commissioner appointed by the President of India.
2 This restriction has been partly lifted with effect from 1 January 2011.
3 Two major hill tribal ethnic groups of Manipur – Nagas and Kukis – occupying the hills areas want bifurcation of the state into two parts – hill and valley, leaving the valley portion, covering 10 per cent of the total state area for Meiteis. The remaining 90 per cent of the hill region is preferred to get further bifurcated into two – one each for Kukis and Nagas.
4 Although they want to promote Kokborok, they encounter difficulty since it does not have a script of its own, and hence its speakers write it in both Bengali and Roman scripts.
5 Please refer to *Indian Streams Research Journal*, Accessed: http://isrj.org/ArticleFullText.aspx?ArticleID=4538.

6 It led to the demise of united Pakistan, hence East Pakistan turned into an independent nation of Bangladesh.
7 As Chief Minister of Assam Mr Tarun Gogoi told journalist Karan Thapar in *Devil's Advocate* programme on CNN-IBN. (Please refer http://www.dailymail.co.uk/indiahome/indianews/article-2200792/Assam-CM-Gogois-illiteracy-jab-state-Muslims.html.)
8 This section is compiled from the writings and inputs of R. N. Ravi, former Special Director, Intelligence Bureau of India.

References

Banton, Michael (1998). *Racial Theories*, 2nd ed., Cambridge: Cambridge University Press.

Baqai, H. (2004). 'Role of Ethnicity in the Conflict Spectrum of South Asia'. *Pakistan Horizon*, 57(4): 57–68.

Berezin, M. (1999). 'Political Belonging: Emotion, Nation, and Identity in Fascist Italy', In Steinmetz, G. (ed.), *State/Culture: State Formation after the Cultural Turn*, Ithaca and London: Cornell University Press, pp. 358–59.

Brass, P. R. (1991). *Ethnicity and Nationalism: Theory and Competition*, New Delhi: Sage Publications.

Cline, L. E. (2006). 'The Insurgency Environment in Northeast India'. *Small Wars and Insurgencies*, 17(2): 126–47.

Conteh-Morgan, E. (2004). *Collective Political Violence: An Introduction to the Theories and Cases of Violent Conflict*, New York and London: Routledge.

Das, S. K. (2007). *Conflict and Peace in India's Northeast: The Role of the Civil Society*, Washington, DC: East-West Centre.

Dale, P. F. and McLaughlin, J. D. (1999). *Land Administration*, New Delhi: Oxford University Press.

Deori, N. (2013). 'Ethnic Fratricide and the Autonomous Councils of Assam', *Yojana*, August 1 (Online Version). http://yojana.gov.in/fatricide.asp.

Dudková, B. E. (2013). 'Consociational Practices as a Conflict Regulation Strategy: A Comparative Study of Ethno-nationalist Movements in Northeast India'. Unpublished PhD Thesis, Department of Social Studies, Masaryk University, Brno: Czech Republic. Accessed on 20 August 2013: http://is.muni.cz/th/274014/fss_m/Dudkova_Thesis_Final.pdf.

Dutta, N. (2012). 'The Myth of the Bangladeshi and Violence in Assam'. Retrieved from http://kafila.org/2012/08/16/the-myth-of-the-bangladeshi-and-violence-in-assam-nilim-dutta/.

Esman, J. M. (1975). 'Communal Conflicts in Southeast Asia', In Glilzcr, N. and Moynihan, D. P. (eds), *Ethnicity: Theory and Experience*, Cambridge: Harvard University Press, pp. 391–419.

Ganguly, R. (2009). *Ethnic Conflict: Causes of Ethnic Conflict*, London: Sage Publication.

Giddens, A. (2006). *Sociology*, 5th ed., Cambridge: Polity Press.

GoA (1976). 'Report of the Employment Review Committee', Guwahati (Assam): Government of Assam, p. 162.

——— (2012). 'White Paper on Foreigners' Issue', Guwahati: Department of Home and Police, Government of Assam.

Goemans, Hein (2006), 'Bounded Communities: Territoriality, Territorial Attachment, and Conflict', In Miles Kahler and Barbora F. Walter (eds), *Territoriality and Conflict in an Era of Globalization*, New York: Cambridge University Press, pp. 25–61.

Gogoi, D. (2007). 'Resolution of a Sunset Dream: ULFA and the Role of the State', In Saikia, J. (ed.), *Frontier Flames: Northeast India in Turmoil*, Delhi: Viking Publisher, p. 122.

Gohain, H. (2007). 'Chronicles of Violence and Terror: Rise of United Liberation Front of Asom'. *Economic and Political Weekly*, 42(12): 1013–15.

——— (2014). 'A Note on Recent Ethnic Violence in Assam'. *Economic and Political Weekly*, 49(13): 19–22.

Government of India (2005). 'Rail Network in North East Region – Expansion and Investment' (Sixth Report of the Standing Committee on Railways – 2004–05), New Delhi: Lok Sabha Secretariat.

Gurr, T. R. (2002). *Peoples versus States: Minorities at Risk in the New Century*, Washington, DC: United States Institute of Peace Press.

Haokip, G. T. (2012). 'On Ethnicity and Development Imperative: A Case Study of North-East India'. *Asian Ethnicity*, 13(3): 217–28.

Hogg, M. and Abrams, D. (1988). *Social Identifications: A Social Psychology of Intergroup Relations and Group Processes*, London: Routledge.

Horowitz, L. Donald (1985). *Ethnic Groups in Conflict*, Berkeley, Los Angeles, and London: University of California Press.

Hrishikeshan, K. (2002). 'Assam's Agony: The ULFA and Obstacles to Conflict Resolution'. *Faultlines*, 12: 23–38.

Hussain, M. (1987). 'Tribal Movement for Autonomous State in Assam'. *Economic and Political Weekly*, 22(32): 1329–32.

Hussain, W. (2013). 'Use of Force to Tame Rebellions in Northeast India: A Cost–Benefit Analysis', Paper presented at a seminar held at Rajiv Gandhi Institute for Contemporary Studies, Delhi, 1 March 2013.

Hutnik, N. (1991). *Ethnic Minority Identity: A Social Psychological Perspective*, Oxford: Cambridge University Press.

Kumar, B. B. (2006). 'Introduction', In Kumar, B. B. (ed.), *Illegal Migration from Bangladesh*, New Delhi: Concept Publishing House, pp. 1–14.

Lacina, B. (2009). 'The Problem of Political Stability in Northeast India: Local Ethnic Autocracy and the Rule of Law'. *Asian Survey*, 49(6): 998–1020.

Mahanta, N. G. (2012). 'The United Liberation Front of Asom (ULFA): Liberating Force or to Be Liberated?' In Arpita, A. (ed.), *Non-State Armed Groups in South Asia: A Preliminary Structured Focused Comparison*, New Delhi: Pentagon Security International, p. 105.

MHA (2013). Annual Report 2012–13: Ministry of Home Affairs. New Delhi: Government of India.
Ngaihte, T. (2013). 'Manipur and Its Demand for Internal Autonomy'. *Economic and Political Weekly*, 48(16): 20–21.
Olzak, S. and Nagel, J. (1986). *Competitive Ethnic Relations*, Orlando, FL: Academic Press.
Penrose, J. (2002). 'Nations, States and Homelands, Territory and Territoriality in Nationalistic Thought'. *Nation and Nationalism*, 8(3): 277–97.
Rammohan, E. N. (2002). 'Blue Print for Counterinsurgency in Manipur'. *Faultlines*, 12: 1–22.
Sack, R. D. (1983). 'Human Territoriality: A Theory'. *Annals of the Association of American Geographers*, 73(1): 55–74.
Saikia, Y. (2004). *Fragmented Memories: Struggling to be Tai-Ahoms in India*, Durham: Duke University Press, p. 166.
Sen, A. (2006). *Identity and Violence: The Illusion of Destiny*, London: Allen Lane.
Smith, A. D. (1986). *The Ethnic Origins of Nations*, Oxford: Blackwell.
Sharma, H. S. (2011). 'Conflict and Development in India's North-Eastern State of Manipur'. *The Indian Journal of Social Work*, 72(1): 5–22.
Singh, M. A. (2011). 'Insurgency and Development in India's Northeast States: A Study on Assam', In Rakhee, B. and Sanjay, P. (eds), *Perilous Journey: Debates on Security and Development in Assam*, Delhi: Manohar Publications, pp. 161–84.
Singh, M. A. (2013). 'Demand for New States in India: A Case Study of the Bodos', In Dutta, P.C. and Komol Singha (eds), *Ethnicity, Resources and Institutions for Development of North Eastern States of India*, New Delhi: Akansha Publishing House, pp. 328–46.
Stanvenhagen, R. (1994). 'Reflections in Some Theories of Ethnic Conflicts'. *Journal of Ethnic Development*, 4(1): 34–39.
Suan, H. K. K. (2009). 'Hills-Valley Divide as a Site of Conflict: Emerging Dialogic Space in Manipur', In Baruah, S. (ed.), *Beyond Counter-Insurgency: Breaking the Impasse in Northeast India*, New Delhi: Oxford University Press, pp. 263–89.
Tensuba, K. C. (1993). *Genesis of Indian Tribes: An Approach to the History of Meiteis and Thais*, New Delhi: Inter-India Publications.
Toft, M. D. (2001). 'Indivisible Territory and Ethnic War', Paper No. 01–08, Cambridge, MA: Weather head Centre for International Affairs, Harvard University.
——— (2005). *The Geography of Ethnic Violence: Identity, Interests, and the Indivisibility of Territory*, Princeton: Princeton University Press.
The Assam Tribune (2013). *New Forum to Espouse Separate State Demand*, Guwahati: The Assam Tribune, Thursday, September 5.
Upadhyaya, A. S., Upadhyaya, P. and Yadav, A. K. (2013). 'Interrogating Peace in Meghalaya'. Core Policy Brief No. 03, Cultures of Governance

and Conflict Resolution in Europe and India, Oslo: Peace Research Institute.

United Nations (2012). 'Land and Conflict', United Nations Interagency Framework Team for Preventive Action.

Verghese, B. G. (2012). 'A Naga Settlement', 23 October 2012, Bangalore: Deccan Herald. http://www.deccanherald.com/content/287243/a-naga-settlement.html.

Walter, B. F. (2006). 'Conclusion', In Kahler, M. and Walter, B. F. (eds), *Territoriality and Conflict in an Era of Globalization*, New York: Cambridge University Press, pp. 25–61.

Weiner, M. (1989). 'The Indian Paradox: Violent Social Conflicts and Democratic Politics', In Varshney, A. and Weiner, M. (eds), *The Indian Paradox: Essays in Indian Politics*. New Delhi: Sage Publications, pp. 21–37.

Weir, N. (2012). 'Primordialism, Constructivism, Instrumentalism and Rwanda', Accessed: https://www.academia.edu/1526597/Primordialism_Constructivism_Instrumentalism_and_Rwanda.

Wright, J. (1999). 'Minority Groups, Autonomy, and Self-Determination'. *Oxford Journal of Legal Studies*, 19(4): 605–29.

Part I

Institution, resources and development

Chapter 1

Ethnicisation of space

Development consequence and political response in Northeast India[1]

Rakhee Bhattacharya

Ethnicisation of space[2] in Northeast India

India's Northeast region for a long time has been viewed as single homogeneous administrative, economic and geographical unit despite its plural character and diverse ethnic communities. This idea of homogeneity was perpetually pursued by the Indian administration for functional convenience, but has remained far off from being successful, as its internal diversity and complexity constantly have been posing challenge with rising ethnic aspiration and demand for autonomy within or outside the constitutional framework. Miri (2007) argues that such challenge primarily is due to difference in views between the insiders and outsiders of the region. The outsiders' view according to him imagines Northeast as unitary entity, inhabited vaguely by differentiated 'tribal' people, who seem 'racially distinct' from Indians elsewhere. And the insiders' diverse view caters to tribes, communities, languages, ethnicity, culture and tradition which are termed as 'egocentric predicament' which over different period of time has given rise to various degrees of ethnic aspirations and ethnocentric conflicts in the region (Ibid.: 4).

The roots of such ethnic aspiration as argued by Sharma (2012) went back to the colonial cartographic principle and their economic and strategic imperatives, which had fixed ethnic identities with definite space and territory. Colonised Northeast also had seen the policies of distinction for the people of Northeast from rest of India with emphasis on the concept of preserving 'tribal' identity of the region (Choudhury and Patnaik, 2008). Since the region never had much cultural link and geographical accessibility with other parts of India, it remained isolated as its connection with neighbouring nations was lost. The policy of gradual segregation of the tribes and non-tribes, the hills and plains; segregation of the tribal population, creation of 'non-regulated', 'backward' and 'excluded' areas/tracts was able to break the centuries of continuum and connectedness. Thus, the region was reconfigured with closed borders for strategic causes, and within the region Inner Line Regulation of 1873 was introduced in the name of preserving indigenous culture and tribal areas and to protect them from the external raids. Thus, trade, economy and culture were heavily disrupted. The valleys and foothills were explored for large-scale commercial plantations like tea despite having confrontation with people like Nagas, whose hunting and food-gathering activities were disrupted due to such encroachment upon land. The creation of tea and oil economy by the British by using labourers brought from outside the region had served as an 'enclave economy' amidst the ethno-traditional economic set-up, which could never bring opulence and welfare to the people of entire region. Such colonial practices along with their myths – the myth of race, core–fringe conflict, isolation and their misinterpretation of history and culture deepened the impact of such isolation and finally alienated the region from rest of India (Bhattacharya, 2011).

In the postcolonial period, such flawed policy rather than being contested was manifested with ethnonationalist leadership through a violent space-centric politics and rhetoric of self-identity. According to Sharma (2012), such emerging principle of recognising identity on the basis of a particular territorial space within the context of multi-ethnic landscape of Northeast India has remained both divisive and damaging for future growth. No other region in India is as deeply affected as its Northeast by the Indian state's policy of space-centric political autonomy to various ethnic communities. Ethnicity, over time, in the region (like South Africa), is deployed much more openly and programmatically as a strategy of state control and social engineering (Preben, 1994). Though the post-independent States Reorganisation Commission (1956) has recommended the merger of other political entities in Northeast with Assam, but over the years the reverse had happened and all states barring Tripura, Sikkim and Manipur were curved out from Assam on ethnic line (Sharma, 2012). Thus, in 1963, Nagaland was formed as a separate state, and eventually Meghalaya, Mizoram and Arunachal Pradesh were created and inevitably gave rise to interstate boundary and ethnic homeland disputes. As a large number of communities have been living in the region for centuries together, in the latter part, many of them started being associated with some territories with their dominant presence, and thus making claim to that territory. Such ethnic assertion began first with Naga insurgency in the 1960s and eventually intensified in other parts of the region by the 1980s, most of which are with space-centric issues. Assam, the nodal state of the region, has undergone phenomenal transformation and has witnessed the rise of many more aspiring ethnic groups and their conflicts, several killings and militant activities. Thus, emergence of various movements like Bodoland, Karbi-Anglong, Mising, Tiwas, Rabha, Koch-Rajbanshis and Kamtapuri has spread across the state and beyond. In other states like Nagaland, Nagalim movement and in Manipur, Naga-Kuki and Meitei movements also are largely space-centric conflicts (Ibid.). Growth of such ethnocentric identity politics and subsequent violence has largely undermined the idea of development and economic modernisation, and the region gradually has become an economic burden to free India. The feeling, therefore, has eventually ingrained in Northeast from the pre-independent 'Colonial policies' to post-independent 'New Delhi policies', and the things were never set right, system functioned within the same structure without opening and imparting any progress to its economy. In words of Gokhale (1998), 'administrative apathy and persistent neglect by the Centre and a policy of isolation pursued by both New Delhi and the local agents in the region, has added to economic deterioration.' These historical realities have challenged the region, snatched away its economic sovereignty, stymied its development, disrupted the means of livelihood of its population, increased its isolation from the rest of the country and accumulated grievances. These further threatened and destroyed social fabric of harmony between different communities along with political unrest and economic underdevelopment.

The chapter therefore explores this causal relationship of Northeast India's ethnocentric and space-centric conflict with consequent damage on its economic development, and what has been the political response to this scenario. It then proposes way forward by revisiting the political economy of the region and by envisioning development in trans-regional space.

Impact and consequence on development

Insurgency and economy

Acute ethnicisation and narrow political vision has significantly affected region's economy and development culture. Cluster of eight states (barring Sikkim in some instances) thus has perpetually remained at the periphery of India's development discourse. The region occupies merely 8.74 per cent of India's total land area, shares around 4 per cent of country's population and contributes insignificantly around 3 per cent of India's domestic product. Persistent ethno-based politics, and electoral power of identity along with acute militancy has drifted the region away from a viable sustainable economy to a non-viable dependent economy. Despite being rich in natural resources, it could not provide any economic space for fair market mechanism and indigenous growth. Assam, for example, which produces more than 50 per cent of India's natural oil and gas, 58 per cent of India's and 30 per cent of world's tea, 60 per cent of India's plywood and about 30 per cent of India's jute, faces the problem of underdevelopment (Das, 1982). Profit from Assam's gigantic tea estate had been persistently flowing out of the region and much later also has gone to the coffer of militant outfits like United Liberation Front of Ahom (ULFA). Similar is the story for Assam's other pride – oil and gas sectors. Such extortions have made severe impact on the economy and pushed it almost into 'economic black hole'. The situation in other disturbed states of the region is not different from Assam. In Manipur, such illegal extortion has almost become institutionalised and became the driving force to initiate and run a separate illegal economy. Along such extortion, a large proportion of development funds flow directly or indirectly to the militant groups through a regime of collusion and intimidation. Thus, only a small proportion of the benefits or incomes that may actually be generated among targeted populations through various developmental programmes, eventually remained with these segments, as militant groups tend to appropriate the surpluses through extortion and 'taxation' (Sahni, 2005). 'Sponsored' insurgencies have become an industry, which probably has become an alternative occupation for many youths of this region. Former Prime Minister Deve Gowda once pointed out that, 'Assam was in the forefront of the economic development of the country 100–150 years ago. It was a pioneering State and enterprising entrepreneurs invested in the development of tea plantations, oil, coal mining, forestry, railways and inland waterways. However, in the recent years, investors have shunned these areas, because some of these states turned inward looking, while others have been afflicted by militancy and terrorism. This has set in a vicious cycle of terrorism, discouraging investments and economic development, leading to growing unemployment, which in turn provided recruits to militancy.'[3] Today by all appearances, economic disparities are not only growing between the region and rest of India, but also growing within the different states in the region, and they have heavily disrupted the investment scenario.

Economic consequences

The region according to Bezbaruah (2015) has wealth of natural resources, which was expected to be a sound economic foundation when the planning process started. Most Northeast states were above national average in terms of State Domestic Product (SDP) and per capita income. After 50 years of planning, the region laments that most of its states are almost at the bottom of the ladder, though there has been some positive movements in the past years. He further cites 'Northeast Vision 2020' documents, which describes economic situation of Northeast as 'anarchic'. The intra-state economic disparity is growing in Northeast as some states like Sikkim, Tripura, Mizoram and Arunachal Pradesh to some extend are performing better than Assam, Manipur, Meghalaya and Nagaland (see Table 1.1).

One of the major economic drawbacks in the region is sheer lack of industrialisation. To encounter this, an industrial policy was initiated in the year 1997 and till the year 2004, the achievement can be reflected in Table 1.2.

This industrial policy was revised as North East Industrial and Investment Promotion Policy in 2007 to give a further boost to its economy, though not much has achieved. Lack of opportunity thus became another vital reason adding to underdevelopment and rising poverty ratios. Job and income-creating activities have been stagnant. The government has been the primary job provider, putting a strain on states' finances (Bzbaruah, 2015). In the context poverty ratios, one can see

Table 1.1 Per capita NSDP in Northeast India (2004–05 prices) (amount in Rs)

States	*1980–81*	*1990–91*	*2000–01*	*2010–11*	*2012–13*
Assam	11,386	13,689	14,595	21,793	37,690
Arunachal	10,514	18,131	23,627	34,366	24,198
Manipur	8,631	10,578	13,409	22,867	25,205
Meghalaya	11,838	15,072	20,099	35,191	39,873
Mizoram	NA	NA	NA	36,732	NA
Nagaland	14,174	20,578	28,400	42,511	43,967
Sikkim	11,239	24,103	22,008	66,136	75,137
Tripura	9,200	11,557	21,607	36,826	42,481
Northeast India	10,997	16,244	20,535	37,053	41,222
All India	**9,748**	**13,400**	**19,687**	**36,342**	**39,168**

Source: Author's estimation. Per capita NSDP figures are converted to 2004–05 prices from CSO 1993–94 and 2004–05 series.

Table 1.2 Number of industries, investment and employment generated in Northeast India (1999–2004)

States	*No. of units*	*Investment (Rs crores)*	*Percentage share*	*Direct employment generated (in no.)*
Assam	520	528.19	49.4	12,422
Arunachal	61	441.01	41.32	6,056
Manipur	11	39.86	3.73	577
Meghalaya	34	31.58	2.96	665
Mizoram	46	19.64	1.84	439
Nagaland	4	4.00	0.39	300
Tripura	5	3.00	0.28	250
Northeast India	681	106,728	100.00	20,709

Source: Lok Sabha Archive Section at http://164.100.47.132/LssNew/psearch/QResult14.aspx?qref=43201.

Table 1.3 Incidence of poverty in Northeast India (%)

States	*1993–94*	*1999–2000*	*2004–05*	*2009–10*	*2011–12*
Assam	40.86	36.09	34.4	37.9	31.98
Arunachal	39.35	33.47	31.4	25.9	34.67
Manipur	33.78	28.54	37.9	47.1	36.89
Meghalaya	37.92	33.87	16.1	17.1	11.87
Mizoram	NA	NA	15.4	21.1	20.40
Nagaland	37.92	32.67	8.8	20.9	18.88
Sikkim	39.01	34.44	30.9	13.1	8.19
Tripura	41.43	36.55	40.0	17.4	14.05
All India	**35.97**	**26.10**	**37.2**	**29.8**	**21.92**

Source: Economic Survey 2001–02, Press notes of the Planning Commission.

from Table 1.3 that all the Northeastern states have higher poverty ratios than the national average, despite attempts through centrally sponsored schemes and other measures. Such a scenario according to many is also due to lack of governance and corruption. Thus, according to Das (2006), in every states of the region, a plethora of allegations and complaints have surfaced regarding the misuse of funds and rampant corruption. The ground-level reality also does not indicate any significant and effectual developmental activity. Media reports have indicated that bureaucrats, politicians, bank managers, contractors and their cohorts siphon-off major portions of the funds, consequently leaving very little – and at times nothing – for the poor. Another report of a committee on Integrated Rural Development Programme (IRDP) in the region mentioned 'an upsurge in insurgency in Assam' as responsible for 'practically no developmental activities in rural areas' and a simultaneous 'flight of capital from the rural areas to urban areas as the former were less secured than the latter'. Such an observation would also indicate that there has been an added impoverishment of the rural areas and a greater lag *vis-à-vis* developmental activity. The general belief that insurgency is fuelled by economic backwardness, therefore, has some basis.[4]

Looking further at other development indicator like credit–deposit ratios in Northeast India (see Table 1.4), it can be seen that the ratios are much lower in the region compared to all India ratio. This very clearly signifies an out-migration of capital from this region due to a severe lack of economic development. Credit–deposit ratio can be treated as the proxy variable for investment scenario in all these states, which reflects an utterly poor state.

Thus, some of these major economic indicators show that development *per se* in some states of this region is far below the national average and one can correlate such under performance with the rising issue of intra-ethnic conflicts and persistent violence through militancy, giving space for identity politics, fragmentation and disparity.

Table 1.4 Credit–deposit ratio in Northeast India

States	*1997*	*2001*	*2005*	*2010*
Assam	41.5	38.1	41.9	40.5
Arunachal	17.6	22.1	30.0	34.4
Manipur	63.4	40.7	42.6	44.8
Meghalaya	14.1	17.3	43.6	32.7
Mizoram	14.7	29.0	59.1	57.7
Nagaland	29.9	13.6	23.2	40.2
Sikkim	18.1	21.7	29.0	31.6
Tripura	39.3	14.5	29.3	49.5
All India	**56.8**	**56.7**	**66.0**	**73.3**

Source: Respective years' Quarterly Statistics on Deposit and Credit of Scheduled Commercial Banks, RBI.

Economic deals

Government of India cannot solely be blamed for always neglecting the region. There has been a perpetual financial support under 'Special Category State'. This cluster of states has been receiving financial assistance from central coffer and its development plans are almost entirely centrally financed on the basis of 90:10 ratio as grant and loan, respectively (see Table 1.5).

As rightly pointed out by Bezbaruah (2015), each union ministry is supposed to spend about 10 per cent of its budget in Northeast. Thus, for example, for the year 2014–15, out of total central sector fund of Rs 484,532 crores, about Rs 48,000 crores was available for the region, in addition to their states' own plan allocation. But so far the pool of unspent funds has accumulated to Rs 15,000 crores. Therefore, any development strategy for Northeast needs to start with an incisive inquiry into why the region could not spend its earmarked financial allocation. Also around 20 centrally sponsored plan schemes are presently operating in the region and per capita fund disbursal under all such schemes are given in Table 1.6.

1 Swarnjayanti Gram Swarozgar Yojana (SGSY)
2 Indira Awaas Yojana (IAY)
3 Pradhan Mantri Gram Sadak Yojana (PMGSY)
4 Special Accelerated Road Development Programme in North East (SARDP-NE)

Table 1.5 Per capita fund disbursement from centre to Northeast India (Rs)

State	*1998–99*	*2000–01*	*2005–06*	*2010–11*	*2012–13*
Arunachal	8,113	8,039	12,934	34,188	40,752
Assam	1,194	1,402	2,579	4,834	8,395
Manipur	3,826	4,193	8,881	17,687	24,463
Meghalaya	3,201	4,038	5,486	12,983	21,674
Mizoram	8,176	8,797	14,817	27,350	NA
Nagaland	5,197	5,941	11,410	23,511	28,994
Sikkim	7,857	9,544	14,431	26,775	NA
Tripura	3,612	4,431	7,858	12,265	16,541
All states	**651**	**873**	**1,555**	**3,277**	**7,047**

Source: Finance Accounts and RBI for Tax and Grant data for respective years. Per capita has been estimated with population for respective years. Population figures are from Central Statistical Organisation (CSO) for respective years.

Table 1.6 Per capita grant for centrally sponsored schemes in Northeast India (Rs)

State	*1998–99*	*2000–01*	*2005–06*	*2010–11*	*2012–13*
Arunachal	470	583	855	2,765	4,729
Assam	70	84	184	440	889
Manipur	222	239	521	1,490	1,133
Meghalaya	214	264	482	1,207	2,903
Mizoram	674	634	910	4,190	NA
Nagaland	217	395	959	0.00	0.00
Sikkim	793	888	1,679	2,403	NA
Tripura	337	281	414	777	1,119
All states	**65**	**68**	**120**	**280**	**683**

Source: Finance Accounts and RBI Grant data for respective years. Per capita has been estimated with population for respective years. Population figures are from CSO for respective years. Nagaland has not received the grant in 2010–11 and 2012–13.

5 The Ministry of DoNER's Plan Scheme of Capacity Building and Technical Assistance
6 Mid-Day Meal Scheme (MDM)
7 Integrated Child Development Services (ICDS)
8 National Rural Drinking Water Programme (NRDWP)
9 Mahatma Gandhi National Rural Employment Guarantee Scheme (MGNREGS)
10 National Rural Health Mission (NRHM)
11 Rajiv Gandhi Grameena Vidyutikaran Yojana (RGGVY)
12 Swarna Jayanti Shahari Rozgar Yojana (SJSRY)
13 The National Social Assistance Programme (NSAP)
14 Kishori Shakti Yojana (KSY)
15 Janani Suraksha Yojana
16 Bharat Nirman Programme
17 Total Sanitation Campaign
18 Scheme of post-Matric Scholarship SC/ST/OBC
19 National Bamboo Mission Northeast
20 Food Security Act

Another concrete policy step towards development of this region was taken up during Mrs Indira Gandhi's regime with the creation of North Eastern Council (NEC) in 1972. The NEC was created by an Act of Parliament, the North Eastern Council Act, 1971, exclusively for regional planning and development. The constitution of the council has marked the beginning of a new era of concerted planned endeavour for the rapid development of the region. Over the last 38 years, NEC has been instrumental as a nodal agency in setting in motion new economic endeavours aimed at removing the basic handicaps that stood in the way of economic development of the region. The broad areas that NEC started functioning in were infrastructure, transport and communication, power, agriculture and capacity building. Sector-wise investment by NEC for three years is given in Table 1.7. A separate union ministry of Development of North Eastern Region (DoNER) also was set up in September 2001 to act as the nodal department/agency of the central government to deal with matters pertaining to socio-economic development of the eight states of Northeast.

Along with central government initiatives, some of the people's movements also have led to various accords and pacts for region's development programme. The movement led by Assam Gana Parishad (AGP) in Assam, for example, has made achievement by making economic deals with Government of India and has culminated to the historic 'Assam Accord', signed in 1985, where Government of India had made certain commitments to render some 'development package' to the state. Some of these are mentioned bellow.

- Numaligarh refinery popularly known as 'Assam Accord Refinery' has been set up as grass-root refinery in the district of Golaghat (Assam) to fulfil the commitment made by the Government of India in 'Assam Accord', at an approved cost of Rs 2,724 crores (Report of Ministry of Petroleum and Natural Gas, 2004–05). This was a part of Government of India's economic package offered for providing required thrust towards speedy economic development of Assam.
- To secure the international border between Assam and Bangladesh and to prevent infiltration of foreign nationals, which is one of the primary issues of Assam movements, construction of Indo-Bangladesh Border (IBB) Road and Fence in Assam was taken up from 1986 to 87 as a 100 per cent centrally assisted project under clauses 9.1 and 9.2 of 'Assam Accord'. The objective of the project was to prevent illegal infiltration through physical barriers like construction of all-weather road and providing barbed wire fencing along the entire stretch of the border to facilitate effective patrolling by the security forces on land as well as river-line routes.
- Under the 'Assam Accord', Government of India made a commitment to revive the 'Jogighopa unit' (mill) of Ashok Paper Mills

Table 1.7 Sector-wise release of funds from NEC (Rs in crore)

Sectors	*2007–08*		*2008–09*		*2009–10*		*2010–11*		*2011–12*	
	Outlay	*Expenditure*	*Outlay*	*Expenditure*	*Outlay*	*Expenditure*	*Outlay*	*Expenditure*	*Outlay*	*Expenditure*
Agriculture and allied sectors	10.42	10.34	12.00	5.40	20.95	20.93	46.75	45.45	69.93	66.15
Water and power development etc.	84.85	91.30	90.45	108.30	147.15	146.53	128.50	112.39	115.53	114.50
Industries/tourism/ minerals	13.08	14.98	11.80	7.70	21.87	21.77	27.71	21.56	23.84	23.21
Transport and communication	404.03	393.44	430.50	430.54	400.23	369.06	391.88	391.93	361.06	330.26
Medical and health sector	25.20	21.24	24.50	21.78	22.00	23.14	38.00	44.95	43.17	42.17
Manpower development	21.00	14.13	30.60	15.70	22.56	24.40	45.90	40.70	62.81	56.89
Externally aided project	23.50	23.50	2.00	2.25	0.00	0.00	0.00	0.00	0.00	0.00
Science and technology	9.81	10.22	15.45	15.87	11.68	11.68	15.05	14.96	16.12	15.88
Information, publicity and public relations	2.60	2.57	3.50	2.34	4.90	2.29	3.60	5.42	6.02	5.86
Evaluation and monitoring	5.51	1.88	4.20	1.60	3.25	1.18	2.61	1.23	1.70	1.68
TOTAL	600.00	583.61	624.00	611.52	624.00	621.00	700.00	678.62	700.00	566.61
% of release		97.27		98.00		99.67		96.94		93.80

Source: Annual Report 2012–13, Ministry of DoNER, http://mdoner.gov.in/sites/default/files/Augumenting%20Infrastructure%2013.pdf.

Limited, a joint venture of the Governments of Assam and Bihar. The Government of India, as per its commitment made under the 'Assam Accord', sanctioned (26 March 1990) Rs 67.08 crore as rehabilitation package for revival of the mill. Accordingly, on 24 September 1990, the mill was taken over by Government of Assam by virtue of the 'Jogighopa (Assam) unit of Ashok Paper Mills Limited Act, 1990'.
- 'Assam Accord' also looked into the human development in the state by creating two central universities, one at Tezpur and the other at Silchar, and most significantly an IIT (Indian Institute of Technology) has been set up at Guwahati.

Despite such initiatives, the region in general had to remain in the domain of low development and lack of modernisation; and factors like ethnicisation, politics of identity, corruption, siphoning of money, lack of good governance and insurgent activities still remain predominant factors for such dismal scenario.

Political response and emergence of regional parties

Lack of deep understanding of regional complexities by Indian national political parties during the post-independent period had also provided enough space to the people of Northeast to feel puzzled, look for a separate political space through movements and revolts and create various regional parties as response to their identity contestation. Assam which was connected to mainland political and freedom struggle discourse since 1885 through Indian National Congress, and later through other major political parties like Communist Party of India (1925) and Bharatiya Janata Party (1980) had witnessed tremendous wave of regionalism in the 1980s on the issue of exploitation and infiltration and a regional party like Asom Gana Parishad (AGP) was formed with massive support from the people of Assam. Thus, Indian National Congress, which was formed in the state with the ideology of social democracy and has ruled Assam since independence till 1978, could find no space in the formation of such regional party. Militant outfit ULFA and its anti-India propaganda had added to negative regionalism and had aggravated the security scenario of the state; and on the same issue Arbinda Rajkhowa, the former leader of ULFA once remarked that

> Twenty seven years ago at the moment of birth of the United Liberation Front of Asom, the prevailing politico-economic and social

> situation was analyzed in detail and we came to the conclusion that the colonial rule and the exploitation thereof was the main factor in pushing the Assamese nation into oblivion. We also arrived at the realisation that apart from the Assamese people the entire indigenous peoples of the region were facing the same fate under Indian colonial rule. Therefore we realised that all the indigenous peoples need to struggle together and that only a united struggle can achieve victory was evident then and now as well. Whatever the Indian media publishes or the local politicians or intellectuals in the Indian pocket say, the undoubted truth is that the survival of the Assam nation is the core issue in the conflict here. At one pole are the indigenous peoples of Assam and on the other is the Indian colonial rule. This is what we refer to as the Assam–India conflict.
>
> (Rajkhowa, 2007)

Such a mindset of anti-India eventually made Assam a cauldron of ethnic tensions, which found space for formation of different regional parties. The several other regional parties that were formed have manifested against Assamese domination. In the year 2005, All India United Democratic Front was formed in Assam by different suppressed sections and linguistic minority groups to protect their democratic rights. Since then Assam has witnessed many more regional parties like Bodo Peoples' Front, United Minorities Front, Barak Valley Territorial Demand Committee, United Tribal Nationalist Liberation Front, some of which grew from armed struggle to parliamentary democracy; and parliamentary arithmetic has enabled many of these regional parties to cling on central power by playing ethnic and linguistic chauvinism. Along with the rise of several regional parities, a persistent anti-India propaganda by the leaders of various militant outfits in different places (some of which were not turned into political parties) also has changed the mindset and political response in the region. It is being commonly felt that national political parties and their vision have failed to respond to regional needs and aspirations of the people, and local urge has led to the formation of various regional political parties for articulation of aspirations by a specific ethnic community or group of communities. Such ethnic dimension therefore has remained one of the most important components of Northeast India's regionalism as almost all regional outfits and political organisations stand for preservation of ethnic identity of one community or the other. Emergence of political parties normally helps to create democracy by enabling the masses to participate in political life. Such democracy is likely to be strengthened in ethnically diverse societies with broad-based, aggregative and multi-ethnic political parties. But in Northeast India such a strategy has been perpetually alienating the region due to fragmented, personalised or ethnically based party system and naturally encouraging continuing conflicts. This never could resolve their fundamental contestation on identity, and such operation of democracy has remained problematic as each group pursues politicisation of identity issues as centrifugal force.

Way back in 1960, All Party Hill Leaders Conference at Shillong was formed as an outcome of ethnic consideration against the imposition of Assamese language as an official state language, and its refusal of the demand for the use of the tribal languages in administration is a pointer towards a strong reaction in the tribal districts and set a trend of ethnicisation of politics in India's Northeast. Thus, various political parties in hill areas merged into All Party Hill Leaders Conference and demanded a separate state within Indian union. In other words, regional political parties in India's Northeast compete within the region with very little scope of extending their scope and ideology beyond 'regional aspirations'. In 1971, Hill State People's Democratic Party was formed in the newly born state of Meghalaya to define the 'law of the land of the Khasi-Jaintia and to fight against infiltration'. Similarly, in other states, regional parties have their own histories and have emerged out of ethnic considerations. In Nagaland, way back in 1963 with the state formation, Naga Nationalist Organisation was formed to govern the state under the leadership of P. Shilu Ao. Naga Nationalist Democratic Party, United Democratic Front, Naga National Democratic Party and United Democratic Front Progressive have been ruling the state in different periods. Recently, Nagaland Peoples' Front got its name changed to Naga Peoples' Front ostensibly to spread its wings beyond administrative and political boundaries – in 'Naga'-inhabited areas of Manipur and Arunachal Pradesh. In Mizoram, Mizo National Front was initiated in 1961, which was initially formed as a sociocultural organisation, Mizo National Famine Front, went on to become an illegal group waging war against India. In 1985, after Mizo Accord, the party was fully integrated into mainstream politics, and now the party plays a significant role in the politics of Mizoram. Parties like Mizoram People's Conference in 1975 were formed on the same issue. Mizo National Front entered into mainstream politics and formed a government after 20 years of struggle in support of Mizoram's independence or separation from India. Even in Manipur, Manipur Peoples' Party was born in 1968 out of ethnic considerations by a group of dissidents from Indian National Congress. The state eventually has seen formation of many more parties and Manipur

Democratic People's Front in 2009 to protect the territorial integrity of the state and to achieve political freedom to decide the future discourse of Manipur. A small party like Kuki National Assembly was formed way back in 1946 to cater to the optical interest of Kuki community in post-independent India. In the multi-ethnic state like Arunachal Pradesh, the People's Party of Arunachal Pradesh was formed in 1977, which was committed to meet the aspiration of the indigenous people of Arunachal Pradesh and to preserve distinctive culture of its varied tribes. Arunachal Congress, the second regional political party was a product of the increased ethnopolitical consciousness of the indigenous people over the issue of the settlement and eviction of refugees. In the state like Tripura though the national parties like Indian National Congress and Communist Party of India have been ruling the parties, yet tribal communities of Tripura are politically conscious and united under the banner of Tripura Upajati Juba Samiti. The parties like Indigenous People's Party of Tripura and Indigenous People's Front of Tripura also fight for their rights and justice and hope for a tribal state. In the year 2002, both these parties have merged to form Indigenous Nationalist Party of Tripura to unite all tribal nationalist forces in a single party. In Sikkim, success of Sikkim Sangram Parishad initiated in 1984 as a regional party is phenomenal. Sikkim Democratic Front was formed in 1993 but with the objective to abide by Indian constitution and to provide all sorts of advantages to the people.[5] It is interesting to observe that all these states except Sikkim had President's Rule between the period from the 1970s to the 1990s, with maximum times in Manipur (seven times), two times in Assam and one time each in rest of the states.

Thus, various such parties were formed at different time periods primarily as a response to war on identity contestation and space-centric politics, which has encouraged ethnic divergence in the region and aggravated intra-ethnic violence and made the region perpetually disturbed and politically unstable. Such a situation also left aside other vital agenda like economy and development and has forced Indian administration to continue to treat the region as one homogeneous backward unit. No path-breaking vision for development was ever imagined for the region until the Look East Policy was introduced in 1991. Look East Policy became a pioneering idea for Northeast to look for a change in the development paradigm. There also happened a gradual change in internal political consciousness and consensus in the region. Thus, only in 2013, ten regional political parties, for the first time, have formed a new political front called North East Regional Political Front for 2014 Lok Sabha Polls. The regional parties that have joined this new party are

Assam Gana Parishad (AGP) of Assam, Naga People's Front (NPF) of Nagaland, United Democratic Party (UDP) and Hill State Democratic Party (HSDP) of Meghalaya, Manipur People's Party (MPP), Manipur State Congress Party (MSCP) and Manipur Democratic People's Front (MDPF) of Manipur, People's Party of Arunachal (PPA) of Arunachal Pradesh, Indigenous People's Front of Tripura (IPFT) of Tripura and Mizo National Front (MNF) of Mizoram to safeguard territorial, cultural, political and economic rights of the people of the region and to continuously strive to protect the distinctive identities of the ethnic tribes and all the people of the region. It also voices concerns of cross-border infiltration, scrapping of Armed Forces (Special Powers) Act, China's initiative to build massive dams in the upper reaches of the Brahmaputra River and implementation of the Look East Policy. This newly formed front also has demanded a comprehensive amendment of the Seventh Schedule of the Constitution with devolution of powers to the states in all matters, except those pertaining to defence, external relations, currency and external trade, and asked for scraping of the concurrent list and to transfer all the subjects in the state list (*Meghalaya Times*, 2014). Such a common regional political front can be taken as a sign of ethnic convergence in the region, and if regional issues are fought together, Northeast India may see a ray of hope with political stability in the coming years. Political and ethnic divergence and rivalry has so far extensively damaged the region's development scenario.

Way forward

It is important to find more ways in the policy debate for ethnic convergence and inclusionary measure beyond political forum. Thus, as argued by Akerlof and Kranton (2000), economics of identity can be inclusionary and can give voice to people to resolve the problem of development deficit of a particular region or a group of people. Northeast India possibly needs such paradigm shift to rewrite its new chapter, where such multi-ethnic landscape can find an alternative space for convergence on a common interest, which may bring a long-term solution to the region. Identity is useful to economists because it suggests a natural way in which behaviour can vary within a population. Since people think about themselves this way, identity corresponds to their own self-classification and also to their classification of others. The combination of identity, social category, norm and ideal allows parsimonious modelling of how utility functions change as people adopt different mental frames of themselves – that is, as they take on different possible identities. Identity describes one special way in which people frame their situation and behaviour (Ibid.). In Northeast, decades-long brutal killings of innocent and poverty-stricken people on identity issue have significantly destabilised regions' economy and development. Today, according to Kotwal (2001), people want developmental measures, employment and not bloodshed, which has been continuing for over two decades. They have fully understood the cumulative disastrous effect of militancy, which is multiplying unemployment, impairing productivity, devastating infrastructure and furthering economic deprivation of the poorer sections of people. A fate they did not deserve, but for the misadventure of a set of misguided youth, kept hostage to the personal benefit of the insurgents (Ibid.). To set things right, it is imperative that some fresh ideas like economics of identity need to be explored with political re-visioning and convergence to redress the legitimate grievances of the people. It is also time to revisit and question the existing political economy of the region to address such multistage problem and the paradox as well as the gap between economic potential and performance in the region. Time is of essence here, and urgency is the need of the hour to contain and stop ethnic conflicts and restore economic links across the border to regain people's confidence. The whole of Northeast India for that matter can rightly be claimed as 'embedded' between the two fastest growing economies of the world – India and China. Also, Northeast India, Southwest China and Southeast Asia can be a 'triangular economic zone' and can bring prosperity to this region through appropriate vision and initiatives. Only with an approach towards integration rather than fragmentation and division, the region can actually transform itself to become the 'geographical heartland' of an emerging economy of Northeast India.

Acknowledgement

I am thankful to Junty Sharma Pathak, Research Associate, Rajiv Gandhi Institute for Contemporary Studies, New Delhi, for assisting me in data compilation.

Notes

1 The views expressed are solely author's and not necessarily of her institute.

2 The phrase – ethnicisation of space – is taken from Kumar Sharma (2012), 'The State and Ethnicisation of Space in Northeast India', In Mahanta, N.G. and D. Gogoi (eds.), *Shifting Terrain: Conflict Dynamics in North East India*, DVS Publishers, Guwahati.

3 As cited by Das (2006); it was quoted from the speech of the former Prime Minister Deve Gowda while announcing the famous package for the region after his seven-day tour of the region in a press statement on 27 October 1996.
4 Cited by Das (2006) from the report of the Committee on Credit Related Issues under IRDP (GSY) in North-Eastern states, Ministry of Rural Development, Government of India, 2000.
5 For details, please refer to Appendix 1.1.

References

Akerlof, George A. and R.E. Kranton (2000). 'Economics and Identity', *Quarterly Journal of Economics*, 115(3): 715–53.

Bezbaruah, M.P. (2015). 'Make in the Northeast', 17 February, *The Indian Express*.

Bhattacharya, Rakhee (2011). *Development Disparities in Northeast India*, Cambridge University Press, New Delhi.

Choudhury, S.K. and S.M. Patnaik (2008) (eds.). *Indian Tribes and the Mainstream*, Rawat Publications, New Delhi, pp 2.

Das, A.K. (1982). *Assam's Agony: A Socio-Economic and Political Analysis*, Lancers Publishers, Guwahati.

Das, H.N. (2006). 'Insurgency and Development: The Assam Experience', accessed 12 December 2013: http://www.satp.org/satporgtp/publication/faultlines/volume10/Article4.htm.

Gokhale, Nitin A. (1998). *The Hot Brew: The Assam Tea Industry's Most Turbulent Decade (1987–1997)*, Spectrum Publication, Guwahati.

Kotwal, Dinesh (2001). 'The Contours of Assam Insurgency', *Strategic Analysis*, 24(12): 2219–33.

Meghalaya Times (2014). 'LS Poll – NE's 10 Regional Political Parties form New Front', 13 January, Shillong, Meghalaya.

Rajkhowa, Arabinda (2007). 'India Committing Genocide in Assam', *The Telegraph*, 30 October, accessed http://telegraphnepal.com/news_det.php?news_id=1362.

Miri, Mrinal (2007). 'Two Views of the North-East', In B.B. Kumar (ed.), *Problems of Ethnicity in the North-East India*, Concept Publishing Company, New Delhi, pp. 3–8.

Preben, Kaarsholm (1994). 'The Ethnicisation of Politics and the Politicisation of Ethnicity – Culture and Political Development in South Africa', *European Journal of Development Research*, 6(2): 33–4.

Sharma, Chandan Kumar (2012). 'The State and Ethnicisation of Space in Northeast India', In N.G. Mahanta and D. Gogoi (eds.), *Shifting Terrain: Conflict Dynamics in North East India*, DVS Publishers, Guwahati.

Sahni, Ajai (2005). 'The Terrorist Economy in India's North East: Preliminary Exploration', accessed 15 December 2013: http://www.manipuronline.com/Features/November2005/northeasteconomy04_1.htm.

Appendix

Table A1.1 Evolution of regional political parties in Northeast India

Assam

No.	*Party name*	*Year*	*Mandate/objective/ideology*
State party			
1	Asom Gana Parishad	1985	It is a direct product of Assam movement against settlement of foreigners and exploitation of Assam's natural resources by the 'outsiders'. Ideology: regionalism and nationalism
2	All India United Democratic Front	2005	Ideology: national inclusiveness It was formed by different suppressed sections of society, underprivileged religious and linguistic minority groups in Assam and aimed at protecting their democratic rights and to empower these segments.
3	Bodoland Peoples Front		Ideology: to administer the Bodoland and Assam as a whole through the principles of democracy, socialism and secularism. To work for strengthening the Indian Nationalism providing due respect to the identities of all sections of people.
Registered (unrecognised) parties			
1	Asom Bharatiya Janata Party	2011	It is a breakaway group of the Bharatiya Janata Party in Assam. Asom BJP was founded by senior BJP leader and former Inspector General of Police, Hiranya Bhattacharya, in 2001. Bhattacharya had objected to the decision of BJP to align with the AGP.
2	Asom Gana Parishad (Pragatishel)	2005	Formed by Prafulla Kumar Mahanta after he was expelled in 2005 by the AGP for anti-party activities. In 2008, the party merged again with AGP.
3	Asom Gana Sangram Parishad	1999	AGSP was launched by the Asom Jatiyatabadi Yuva Chatra Parishad (AJYCP) in 1999. In the 2001 state legislative assembly polls, AGSP joined the Rashtriya Democratic Alliance led by the Nationalist Congress Party.
4	Asom Jatiya Sanmilan	1998	It was founded by AGP dissident Bhrigu Phukan in 1998.

(*Continued*)

Table A1.1 (Continued)

No.	*Party name*	*Year*	*Mandate/objective/ideology*
5	Autonomous State Demand Committee	1991	Autonomous State Demand Committee, originally the Peoples Democratic Front, was set up as a mass organisation of the Communist Party of India (Marxist–Leninist) Liberation in order to fight for statehood for the Karbi Anlong region in Assam.
6	Bodo People's Progressive Front	2005	It is a political party formed for participation in the Bodoland Territorial Areas District elections.
7	Cachar Congress	2011	A regional political party in the Barak Valley formed ahead of the 2001 Assam assembly elections by a group of Indian National Congress dissidents. The founders of CC were dissatisfied with the distribution of tickets for election and the leadership of Santosh Mohan Dev.
8	Krishak Banuva Panchayat	1940	It functioned as the open mass front for the Revolutionary Communist Party of India in Assam.
9	Marxist Manch		It is a political party which emerged as a local splinter group of Communist Party of India (Marxist).
10	Natun Asom Gana Parishad		NAGP was formed through a split of AGP.
11	Plain Tribal Council of Assam		In 1966, the PTCA launched a militant agitation for a separate tribal state called 'Udayachal'.
12	Rashtriya Democratic Alliance		It was a front of five political parties contesting the 2001 state legislative assembly elections in Assam. Its constituent parties included Nationalist Congress Party, Asom Jatiya Sanmilan, Asom Gana Sangram Parishad, Purbanchaliya Loka Parishad and Janata Dal (Secular)
13	Trinamool Gana Parishad	2000	It was formed as a splinter group of AGP in 2000. The party was led by Atul Bora, then Public Works Department Minister in the Prafulla Mahanta Cabinet.
14	United Minorities Front	1985	It was set up in 1985 by the All Assam Minority Students Union, Citizens Rights Preservation Committee and other religious/linguistic minority people of

No.	Party name	Year	Mandate/objective/ideology
			Assam, as a response to the Assam movement of the All Assam Students Union and the signing of the Assam Accord. Its support base is mainly from Bengali Muslims and Hindus.
15	United People's Party of Assam		It was an ally of AGP and took part in an AGP-led government in the state.
16	Barak Valley Territory Demand Committee		
17	Loko Sanmilon		
18	United Reservation Movement Council Of Assam		
19	United Bodo Nationalist Liberation Front		
20	United People's Front		
21	United Tribal Nationalist Liberation Front	1984	It was launched by the PTCA (Progressive) (splinter group of Plain Tribal Council of Assam) in 1984. It demands statehood for the Bodo areas.

Arunachal Pradesh

State parties

1	People's Party of Arunachal	1977	It was formed and committed to meet aspiration of the indigenous people of the State and to preserve distinctive culture of its varied tribes.

Registered (unrecognised) parties

1	Arunachal Congress	1996	Formed as a splinter group of the Indian National Congress, when the local party leader and Chief Minister Gegong Apang revolted against the then Congress leader P.V. Narasimha Rao. Just before, 2009

(*Continued*)

Table A1.1 (Continued)

No.	*Party name*	*Year*	*Mandate/objective/ideology*
			Assembly election, Arunachal Congress merged with the Indian National Congress. It was a product of the increased ethnopolitical consciousness of the indigenous people over the issue of the settlement and eviction of refugees.
2	Arunachal Congress (Mithi)	1998	A breakaway group of Arunachal Congress. It was formed in 1998 when Mukut Mithi led a revolt against Gegong Apang. It subsequently merged with the Indian National Congress.
3	Congress (Dolo)	2003	A group that split away from the Indian National Congress.

Meghalaya

State parties			
1	Garo National Council	1946	The party fights for the creation of a Garo state, to be carved out of three districts of Meghalaya, West Garo Hills, East Garo Hills and South Garo Hills.
2	National People's Party	2012	P.A. Sangma, a prominent politician, had announced to form a new political party soon after his expulsion from the Nationalist Congress Party in 2012, when he refused to accept party decision to quit the Indian presidential election. It is supposed to be a tribal-centric party.
3	Hill States People's Democratic Party	1971	The very purpose is to defend the 'law of the land of the Khasi-Jaintia' and to fight against any injustice committed against them. To stop, prevent infiltration by foreigners' especially Bangladeshi citizen. To develop a sense of pride among Khasi-Jaintia people and to promote harmony.
4	Meghalaya Nationalist Congress Part	2003	It was formed by Cyprian Sangma when six out of fourteen Nationalist Congress Party legislators broke away. Four of the six were immediately given cabinet berths in the D.D. Lapang government. It subsequently merged with the Indian National Congress.
5	United Democratic Party	1998	

No.	Party name	Year	Mandate/objective/ideology
Registered (unrecognised) parties			
1	Achik National Congress (Democratic)		
2	All Party Hill Leaders Conference	1960	In 1960, various political parties of the hill areas merged into the All Party Hill Leaders Conference and again demanded a separate state. The passage of the Assam Official Language Act, making Assamese the official language of the state, and thus the refusal of the demand for the use of the tribal languages in administration, led to an immediate and strong reaction in the hill districts of Assam.
3	Khasi Farmers Democratic Party		
4	Khun Hynnieutrip National Awakening Movement	2003	
5	Meghalaya Democratic Party	2002	
6	People's Democratic Movement	1998	
7	Hill Peoples Union	1988	
8	Public Demands Implementation Convention	1983	
9	Youth Democratic Front	1993	
Nagaland			
State parties			
1	Naga Peoples' Front		The Nagaland Peoples' Front got its name changed to the Naga Peoples' Front ostensibly to spread its wings beyond administrative and political

(*Continued*)

Table A1.1 (Continued)

No.	Party name	Year	Mandate/objective/ideology
			boundaries in 'Naga'-inhabited areas of Manipur and Arunachal Pradesh. Ideology: regionalism
Registered (unrecognised) parties			
1	Nagaland Democratic Party	1999	On 22 March 2004, it merged into the Nagaland Peoples Front.
2	Naga National Democratic Party	1964	
3	United Naga Democratic Party		
Mizoram			
State parties			
1	Mizo National Front	1961	Born out of long armed struggle espoused the cause of 'subnationalism' joined the mainstream politics after the Mizo Accord. Ideology: populism, Mizo nationalism, social liberalism, democratic socialism, social democracy, Christianity, social populism
2	Mizoram People's Conference	1975	It was formed by Thenphunga Sailo. It is still the third largest party in the state, but is far surpassed by its rivals, the Mizo National Front and Indian National Congress.
3	Zoram Nationalist Party	1997	It was formerly known as Mizo National Front (Nationalist). It is led by former MP Lalduhoma. It was formed in 1997 through a split in the Mizo National Front.
Registered (unrecognised) parties			
1	Ephraim Union		It belongs to the section in Mizoram that claims that the Mizos are descendants of the ten lost tribes of Israel, and it advocates conversion to Judaism. It struggles for making it easier for Mizos to convert to Judaism and migrate to Israel.
2	Hmar Peoples Convention	1986	The group advocates for the rights of members of the Hmar people. The group was formed following the conclusion of peace between Indian

No.	Party name	Year	Mandate/objective/ideology
			government and the Mizo National Front in 1986, which did not take into account the Hmar wish for autonomy. Their main demand is to create an autonomous region in the north and northeast of Mizoram.
3	Lairam People Party	1992	
4	Maraland Democratic Front	1997	The party is active among the Mara people in the southern part of the state.
5	Mizoram Congress Party	2005	

Tripura

Registered (unrecognised) parties

No.	Party name	Year	Mandate/objective/ideology
1	Indigenous Nationalist Party of Twipra	2002	It was formed as a merger of the Indigenous People's Front of Tripura and the Tripura Upajati Juba Samiti in 2002. Its formation was pushed through after pressure from the National Liberation Front of Tripura, which wanted to unite all tribal nationalist forces in a single party.
2	National Conference of Tripura	2006	It comprises many dissidents of the Tribal Students Union, Tribal Youth Federation and Gana Mukti Parishad and also many disgruntled leaders from Indigenous Nationalist Party of Twipra.
3	National Socialist Party of Tripura	2003	It was formed when Hirendra Tripura and others broke away from the Indigenous Nationalist Party of Tripura.
4	Tripura Rajya Muslim Praja Majlish	1946	It competed with Anjuman Islamia over the political influence over the Muslim community, but failed to make any lasting impact.
5	Tripura Ganatantrik Manch		Is a splinter group of Janganotantrik Morcha, which itself is a splinter group of Communist Party of India (Marxist) in Tripura.
6	Indigenous National Party of Tripura		A regional party to address the hopes and aspirations of tribal and was propped by the Congress to tackle the growing left domination.

(Continued)

Table A1.1 (Continued)

No.	*Party name*	*Year*	*Mandate/objective/ideology*
7	Indigenous People's Front of Tripura		
8	Tripura Upajati Juba Samiti		It united the tribal people in fights for their rights and justice.

Sikkim

State parties			
1	Sikkim Democratic Front	1993	Ideology: democratic socialism. According to the leaders of Sikkim Democratic Front, the objectives of the party are to abide by the Indian constitution and law, to maintain integrity, sovereignty, equality of the country. It aims to provide the people of Sikkim all advantages provided by government. It fights for the uplift of educationally, socially and economically backward and underdeveloped people. Political freedom, an all-round prosperity for Sikkimese people and equal opportunity in every sphere activity for the participating Sikkimese are some of the basic goals that constitute this party's agenda.
Registered (unrecognised) parties			
1	Sikkim Gorkha Prajatantric Party		
2	Sikkim Janashakti Party	1997	It was founded in 1997, when Tara Man Rai broke away from Sikkim Ekta Manch. It subsequently merged with the Indian National Congress.
3	Sikkim National Liberation Front		
5	Sikkim Ekta Manch	1997	
6	NEBULA Party	1999	
7	Sikkim Janata Party	1972	Founded by Lal Bahadur Basnet. The party was active in the struggle for democratic reforms. Basnet had been prosecuted by the monarchy in 1966. He had founded this party after that.

No.	Party name	Year	Mandate/objective/ideology
8	Sikkim National Congress	1962	It claimed to represent all ethnic groups of Sikkim. It opposed the monarchy and worked for democratic reforms. After the merger of Sikkim with India, it also merged with the Indian National Congress.
9	Sikkim Sangram Parishad	1984	Nar Bahadur Bhandari from Sikkim Janata Parishad party gained power in Sikkim. In 1984, Bhandari dissolved Sikkim Janata Parishad and formed the Sikkim Sangram Parishad.

Manipur

State parties			
1	Manipur Peoples Party	1968	It was born out of ethnic considerations. It was founded in 1968 by a group of dissidents from the Indian National Congress.
2	National People's Party	2012	P.A. Sangma had announced to form a new political party soon after his expulsion from the Nationalist Congress Party in July 2012, when he refused to accept party decision to quit the Indian presidential election, 2012. It is supposed to be a tribal-centric party.
Registered (unrecognised) parties			
1	Democratic Revolutionary Peoples Party		
2	Manipur State Congress Party	1997	It is a breakaway faction of the Indian National Congress.
3	Federal Party of Manipur	1993	It was born out of ethnic considerations. The party had an average presence in the localities of the state and the initial aim of this party was to uplift the backward people of Manipur irrespective of their caste, creed, colour and sex.
4	Manipur National Conference	2002	It was formed in 2002, when a split occurred in the Manipur State Congress Party.
5	Nikhil Manipuri Mahasabha	1934	

(*Continued*)

Table A1.1 (Continued)

No.	*Party name*	*Year*	*Mandate/objective/ideology*
6	People's Democratic Alliance	2012	It was formed after eleven non-Congress parties came together to form a coalition, in an effort to form a non-Congress government after the assembly polls in 2012.
7	Kuki National Assembly	1946	It was formed with the primary objectives of building a consciousness of common identity, to cater the political interests of the Kuki people in the post-independent era, and creating a single political unit for the Kukis.
8	Manipur People's Democratic Party	2006	
9	Naga National Party		It works among the Naga minority. It favours negotiated settlements of the conflicts in the northeast, unification of Naga groups and maintaining of Naga identity.
10	Manipur Democratic Peoples' Front	2009	

Source: Multiple sources have been used: Election Commission, Lok Sabha, Rajya Sabha, several articles and reports, websites of many parties.

Chapter 2

Beyond ethnicity

Development and reconciliation

Uddipana Goswami

Introduction

India's Northeast region (NER) has often been called its 'troubled periphery', owing largely to the violent conflicts of varying intensity that have characterised the region during much of its postcolonial history. The manifestations of these conflicts have also been varied, ranging from mass civil disobedience movements to armed militancy and secessionist insurgencies. However, one factor that is common to them is that they are all invariably linked to contestations over identity and ethnicity.

A multicultural, poly-ethnic region, ethnicity and the ethnonationalist[1] sentiments of the numerous communities of the region have often lent themselves to – and been used towards – fanning political conflicts. On many occasions, conflicts have broken out when traditional rivalries and existent fissures between communities have deepened in response to expedient demands and clamant considerations. These fault lines and fissures have often been exploited by various quarters to make latent conflicts manifest. Meanwhile, the Government of India (GoI) has failed to address the underlying conflict causes and responded to the ethnic conflicts mainly with large-scale militarisation and a strategy of co-option of the ethnic elites. This has led to the rise of new conflicts and exacerbation of old ones. To boot, a skewed approach towards the development of the rich resources of the region has only deepened the divide between communities.

Given the never-ending cycle of violence that the NER has been plunged into, this chapter argues that any sincere effort at conflict resolution/transformation must now look beyond specific ethnic and ethnonationalist concerns. Even-handed development and ethnic reconciliation, it argues, is the way forward towards building pathways[2] between communities and ushering in sustainable peace.

The instrumentalist nature of identity politics

The 'primordialists' in ethnic studies maintain that 'primordial ties' are at the root of identity formation (Geertz, 1996:40–1); that ethnic identity, like many other social ties, stems from the 'givens' and that the determinants of an ethnic identity – the 'congruities of blood, speech, custom and so on' – have 'an effable, and at times overpowering, coerciveness in and of themselves' (ibid.:41–2). They believe that once formed/constructed, ethnic identities are fixed and immutable (Bayar, 2009:2). Also, these primeval identities then interact with the drive for an 'efficient, dynamic, modern state' (Geertz, 1996:40–1) to give rise to ethnonationalist sentiments.

In contrast, the instrumentalist approach to ethnic and ethnonationalist identity formation treats the phenomenon as a 'social, political and cultural resource for different interest – and status-groups' (Hutchinson and Smith, 1996:8). Ethnic consciousness in recent decades has indeed been the result of post-colonial power struggles 'over new strategic positions of power – places of employment, taxation, funds for development, education, political positions, and so on'. Ethnic formations then have come into being as a kind of 'political grouping within the framework of the new state' (Cohen, 1969:199).

It is not the intention of this chapter to debate or defend the relative merits of either the instrumentalist or the primordial positions. It definitely does not dispute the primordial elements in pre-colonial ethnic formations. It does, however, seek to limit itself to an instrumentalist understanding of the multiple processes of ethnonationalist identity formation and re-formation in NER in the post-colonial period, processes that have subsequently led to the outbreak of the many intractable wars here. It will, therefore, look only at those elements in the construction of these identities that have fuelled conflicts in recent times. It would, for instance, consider the post-colonial Naga ethnonationalist identity in this light; or, for that matter, the various reformulations of the Kuki identity *vis-à-vis* the Naga. In what can be described as ethnic passing,[3] many Kuki tribes, therefore, 'including Anal, Maring, Monsang and Moyon, etc. has been assimilated into the Naga fold' (Kipgen, 2013). Some of these passings happened through coercion and conversion, but there are also many documented instances of ethnic infiltration among communities in the region, or of 'smaller' ethnicities (collectively and consciously) merging (or trying to merge) their identities with 'larger' ones. In Assam and Tripura, for example, detribalisation was seen as a means of entry into the dominant Hindu universe. Francesco Caselli and Wilbur John Coleman II (2006:1) discuss this phenomenon thus:

> Each society is endowed with a set of wealth-creating assets, such as land and mineral resources. There is therefore an incentive for agents to form coalitions to wrest control of these assets from the rest of the population. Once a coalition has won control over the country's riches, however, it faces the task of enforcing the exclusion of non-members. In particular, agents not belonging to the winning coalition will attempt to infiltrate it, so as to participate in the distribution of the spoils.

The ethnic clashes between the communities of the region in recent decades, the Kuki–Naga conflict in Manipur being just one instance, needs to be examined under this instrumentalist light. The search for conflict transformation and ethnic reconciliation also, then, will have to begin with the same understanding.

Ethnic conflict and state response

The problem with ethnic conflicts like the ones raging in the region, however, is that 'concrete and tangible stakes' become infused with 'symbolic and even transcendent qualities'. And because tangible stakes are bestowed with intangible qualities, they become 'difficult to divide' thus encouraging disagreements and 'more hostility' (Vasquez and Valeriano, 2009:194–5).

Territory is one of these tangible stakes, and territorial exclusivity in the form of grant of Sixth Schedule status, creation of autonomous regions, demarcation of Inner Lines and so on, are often fetishised and treated as the answer to all 'intangible' ethnic aspirations[4]: preservation of culture and identity against the onslaught of settler/migrant communities as well as the domination of larger neighbouring autochthons being the primary. The cry for a 'homeland' – and each 'homeland' that is demanded by every ethnic group (or by the same ethnic group at different points in history) may have a different constitutional status – is not just about territorial demarcation; it is inextricably linked to the particular community's threat perception. The 'homeland' is where the traditional and customary survives, it is where the community is safe from 'outsiders' and, of course, it is where its members have entitlements to their own resources. The problem, however, lies in the fact that territory is 'difficult to divide', to quote Vasquez and Valeriano (2009:194–5) once again. Falling as it does in one of the most historically vibrant and busy migration routes of humankind, demographically, the NER has always had a multicultural makeup. It was the British colonisers who introduced to this multicultural map the notions of fixed boundaries and 'settled' frontiers: notions which were retained under the post-colonial regime as well. Before that, as Chatterjee (1994:223) has shown, 'communities were fuzzy'. By instituting new ways of collecting information about colonised populations 'in the form of maps, settlement reports . . . statistical information, censuses' and so on (Das 2007:24),[5] they (the colonisers) introduced a sense of differentiation along ethnic lines. Post-colonial policies and administrative instruments perpetuated this sense of difference further, and ethnic boundaries between communities congealed more and more to reflect not merely the cultural/intangible but also the territorial/tangible.

The problems with territorial demarcation of ethnic communities, though, are many. To begin with, there is no clear demarcation between many of them ethnologically. The processes of ethnic passing and infiltration (including detribalisation) discussed earlier contributed to this complication. For example, many ethnographers agree that the Koch people of Assam and West Bengal are detribalised Bodos. Historically, therefore, they shared the same 'homeland'. However, their ethnic identity-formation movements took divergent turns in the post-colonial period so that their 'homeland' demands now share an overlapping cartography. The Bodos enjoy territorial autonomy over a major portion of this map, and the Koch people continue to protest against this and the conflict between the two communities continues to fester as a result. So whose homeland is it now?

And then there is another facet to ethnic passing/infiltration that complicates the matter further: in many cases, these processes were arrested or stalled at various points in their development trajectory. Like the Koch, the Modahis of Assam, for instance, were also originally members of the Bodo group of Tibeto-Burman communities. Later, they converted to Hinduism. But they remained outside the orbit of caste-Hindu society and were not accepted unconditionally into the Hindu universe because of their inability to give up alcohol (*mod*) altogether. How would one determine the location of this community – and other such in-between communities – in contemporary spatial politics?

The clamour for separate, 'homelands' and territorial autonomy is definitely a post-colonial political phenomenon. After all, migration and population movement has been more the norm than an exception in the region. In fact, certain communities, like the Bodos of Assam, did not entirely take to sedentary lifestyles till as recently as the turn of the nineteenth century (Guha, 2000:34). It was only with the influx (since the nineteenth century) of large numbers of East Bengali and Adivasi peasants from other parts of British India that pressures on land and limited resources increased, forcing them to take to the plough instead of the hoe (ibid.); the hoe being the main instrument of *swidden* or shifting cultivation, the plough of settled. When such a community takes to ethnic cleansing in order to ensure numerical majority in an exclusive 'homeland' (as Bodo militants have done in recent decades), one is again faced with the question, which homeland?

The policymakers of the country (India), however, seem to lack (or ignore) such a nuanced understanding of the relation of ethnicity with territory. So, they refrain from looking for extraterritorial solutions which would address not just the aspirations of one particular (the dominant in many cases) community but also accommodate those of the 'national minorities (who) do not form the majority of the population in areas in which they reside'. One such solution could be the grant of cultural autonomy. Under this arrangement, the community, not the territory they inhabit, is granted autonomy. It allows the community to be governed by its own institution and legislations in the interest of preserving, protecting and promoting its identity (Benedikter, 2007:43–4). The closest the Indian Constitution has come to granting anything like cultural autonomy is in the provision, under the Sixth Schedule, for the formation of autonomous regional councils in an area where 'there are different Scheduled Tribes in an autonomous district'.[6] These regional councils, therefore, would still have the superstructure of the territorial autonomous council above them.

Currently, the Bodos of Assam enjoy a Sixth Schedule status under the Indian Constitution which grants them territorial autonomy. The BTAD spreads over much of western and central Assam. But if the aim of territorial autonomy is to enhance peaceful integration in a democratic system (Benedikter, 2007:62), it has failed to do so in the case of Bodo society which is in the throes of factionalism and fratricidal violence in Assam. A close look at the nature of the autonomous arrangement and the way in which the autonomous council has been functioning in Bodoland strongly suggests that the only segment of the Bodo community that has benefitted from autonomy is the ethnic elite.

Indeed, all over the NER, because of a proper lack of implementation and realisation of the actual intended principles behind the territory-based protective discrimination mechanisms, they have been reduced to mere sops for the ethnic leadership. It is one way by which the Indian State has been able to control the intensity of the otherwise intractable ethnic conflicts in the region. Meanwhile, the aspirations of the larger sections of the community remain unaddressed.

In its search for conflict settlement,[7] the Indian State has always preferred to deal with the ethnic elites – who may be active or former members of the political, armed and/or civil society leadership. They are co-opted with monetary inducements and the promise of participation in power-sharing arrangements after prolonged militarisation and imposition of draconian laws has eroded their resistance. They are then installed in the new power centres – in the context of our current discussion, the territory-based autonomous councils – where they continue perpetuating the same old ethnic hierarchies. Additionally, they now begin acting as the agents of the state, aiding the state in implementing its same old policies of suppressing ethnic aspirations, thus keeping alive the cycle of violence and conflict.

Such cosmetic territorial makeovers – and creation of new centres of power and control – also increase the competition over the available resources in the newly demarcated territories, thus once again, abetting conflicts and escalating violence. Indeed, on various occasions, the Indian State has consciously adopted policies aimed at mobilising and 'developing' these resources that are not only anti-people but also patently conflict inducing. Some of these policies have even brought two or more communities to confrontation with one another. A clear example is the 1,500-MW Tipaimukh Multipurpose Hydroelectric Project proposed to be built over Barak River in Manipur. Besides the enormous environmental cost[8] that will ultimately impact all communities living in the region, many questions about rehabilitation and loss of agricultural land have also been raised ever since the project was initiated. But most importantly, and in relation to our current discussion, it has already incurred a huge cost in terms of peaceful ethnic coexistence.

The dam is to be built in a Hmar-dominated area and some sections of the community have welcomed the project in the hope of being benefitted by it. Vocal protests, however, have been raised by the non-Hmar communities which will be the most adversely affected. Many Zeliangrong villages, for example, will be submerged and leaders of the community have questioned: 'If some people shall get a little benefit at the cost of our people how can the government trade off one community's future against the other' (Sethi, 2006). It has also forced community leaders to threaten armed resistance; Guiliang Panmei, adviser of the Zeliangrong Women's Union, has been quoted as saying: 'If this is what the government wants to do then we shall have no option but to pick up arms. We shall defend our way of life and our lands' (ibid.).

It has been a common experience in the region that when numerically small communities – and many ethnic communities of the NER have very small numbers – find no redress, have no means of getting their demands heard constitutionally or making their presence felt in the democratic political system, they resort to armed guerrilla warfare. To crush these insurgencies, the state immediately resorts to militarisation. The end result of the state and non-state actors thus interfacing is a legitimisation of violence at various levels. In the entire region, this has led to widespread criminalisation of society; violence giving birth to more violence, brutalisation eroding ideologies and State-sanctioned terror giving rise to a disregard for peaceful alternatives (Goswami, 2010). Consequently, what started as insurgency ends up as a law and order problem, a social phenomenon that is beyond ethnic aspirations or ideals.

Development and reconciliation

Any effort at breaking this endless cycle of violence also, then, should begin at this point: beyond ethnicity. Arguably, the state's success in reducing ethnic insurgencies to a law and order problem – which can then be tackled at an entirely different level – is no mean feat, it definitely makes the conflicts easier to 'manage' at a superficial level. The inherent problems, though, remain unaddressed and both latent as well as manifest conflicts continue to fester and be fostered. Conflict as an industry keeps growing and certain constituencies and agencies continue to benefit from it.

The final argument of this chapter is that sustainable conflict transformation[9] and peace building need not be any less rewarding. What is more, these rewards are more equitably distributed. It advocates, therefore, that the state should cease resorting to the conflict settlement approach that is reflected in its concentration on power-sharing mechanisms and skewed 'developmental' policies. Instead, it needs to work out longer-lasting peace-building measures like sharing resources with the communities in conflict.

To illustrate: the GoI has proposed to tap the immense water resources of Arunachal Pradesh by building 168 mega dams in this highly seismically sensitive zone. Once the project is completed, in all likelihood, only a miniscule proportion of the thousands of megawatts of power produced by these dams will be utilised within the region while the rest will be siphoned to the mainland and/or exported to the neighbouring countries. Meanwhile, not only will the dams have destroyed the delicate and relatively untouched ecology of Arunachal, but they will also keep the potential for interstate disputes forever alive. After all, the dams will remain a constant threat to the people downstream, mostly in Assam, huge areas of which will be inundated and thousands of people displaced.

To contrast: in neighbouring Mizoram, an alternative to big dams has been successfully implemented – by the state itself – in at least one instance. At Lohry in Saiha district, a micro-hydel project was completed in 2011 under the National Rural Employment Guarantee Scheme (NREGS).[10] Local villagers provided the labour for the project and they were paid under the scheme. The power generated from this project will also go to these locals (Vanglaini, 2011). Everywhere in the region, there are numerous such water bodies that can be tapped safely for production of energy while generating employment for the local population. If the state undertakes more such developmental measures, it can not only empower the people at the grassroots but also help in erasing the sense of alienation among them. Such policies, after all, infuse an understanding of responsibility and accountability among the local population by giving them a sense of ownership over the implemented projects and the resources they tap.

Considerations of and contestations over ethnic identities need not enter into these micro-level projects that can be replicated in different places dominated by different communities. Even-handed development should be the driving principle. What is more, the state also does not stand to lose anything if the existing government's populist schemes and sponsored programmes (like the NREGS mentioned earlier) succeed in this manner.

Community ownership and accountability aside, it might be further argued that peace building should also be gainful even at an individual level, that there has to be a space reserved for the private sector as well. Indeed, it has been observed that allowing the private sector space and opportunity to develop at the local level has often had a salutary effect in peace building (Killick *et al.*, 2005). Local businesses are after all, a 'part of the existing conflict context', highly 'networked' and

> . . . constituting a powerful section of society (either in terms of political leverage or, at the lower level, the kinds of services provided) with a variety of linkages to different social and political actors and strata, through business relations (with staff, business

partners, etc.), but also along other lines, including the political, cultural, ethnic or religious. (ibid.)

As such, they are strategically positioned to intervene in a variety of ways. The state should, therefore, prop them up and use them in its conflict transformation endeavours. The burgeoning of local businesses and entrepreneurs – and successful ones at that – in recent years in the region proves beyond a doubt that there are a lot of resources in the private sector of the region. These can be tapped towards a peaceful transition into a post-conflict environment. Once again, ethnic considerations need not come into play here, and thereby, further lacerations to the multi-ethnic fabric of the region can also be avoided.

The already existing wounds, of course, need to be healed simultaneously. Initiating a process of ethnic reconciliation[11] is necessary in order to restore 'social harmony of the community in general and of social relationships between conflict parties in particular' (Boege, 2006). Rethinking the existing development paradigm is one way forward in this direction. Journalist and writer, Subir Bhaumik (2007), has held up an illustration of just how this can be done in the case of his home state, Tripura. In post-colonial times, ethnic conflicts have claimed thousands of lives and livelihoods and displaced many in Tripura. One way of effecting ethnic reconciliation here, Bhaumik (ibid.) argues, is to decommission a near-defunct hydel projects to clear land enough to rehabilitate almost the entire landless indigenous population: 'The Tripura government can get up to 65 km^2 of fertile lands that has been under water for thirty years by decommissioning the 10 MW Gumti Hydel project which has stopped producing electricity during summers because of lack of water in reservoir' (ibid.:35). He further suggests that the government should instead concentrate on developing the large reserves of natural gas discovered in Tripura. This would not only assuage hostilities between communities but also benefit the government which can save on the upkeep of an unproductive project and concentrate on developing more profitable resources. A Gumti dam may not be readily available everywhere in the region to lend itself towards decommissioning and thus aiding ethnic reconciliation, but the principles behind this case study may be used as guidelines elsewhere. These include rethinking of the cost effectiveness of existing – and especially conflict-inducing – 'development' projects and re-evaluating their worth, and where required shutting them down or demolishing/abandoning them to serve the ends of even-handed development. This would, of course, postulate a sincere commitment towards ethnic reconciliation. And therein lies the rub.

Notes

1 Ted Gurr (2000:17) defines ethnonationalists as 'indigenous peoples who had durable states of their own prior to conquest, such as Tibetans, or who have given sustained support to modern movements aimed at establishing their own state, such as the Kurds'.
2 In her study of the conflictive and co-operative relations between communities in the Philippines, Melanie Hughes McDermott (2000) has used the concepts of 'boundaries' and 'pathways'. According to such a classification of relations, 'boundaries' establish the differences between peoples, whereas 'pathways' illuminate boundary-crossing social relationships. There are boundaries of identity (of delimitation between groups), territorial boundaries (of homeland or nativity), boundaries of trust (expectations of habitual and amiable interaction with other groups) and boundaries of interaction (exclusive spheres of the social universe).
3 'Although the actors themselves . . . speak as if ethnic boundaries are clear cut and defined for all time; and think of ethnic collectivities as self-reproducing bounded groups, it is also clear that from a dynamic and procedural perspective, there are many precedents of 'passing' and the change of identity, for incorporation and assimilation of new members and for changing the scales and criteria of collective identity' (Tambiah 1989:335–6).
4 In the NER today, we even have dominant, non-indigenous and autochthonous communities crying out for drawing of Inner Lines: these lines, beyond which 'the tribes are left to manage their own affairs' (Chaube, 1973:12–15) were aimed initially at protecting the indigenous communities under threat from non-indigenous settlers.
5 Though Chatterjee and Das as quoted here were discussing caste- and religion-based communities, the same is equally applicable to the ethnic communities of Northeast India as well.
6 Retrieved from www.lawmin.nic.in, 12 February 2014.
7 The conflict settlement approach 'defines conflict as a problem of political order and of the status quo: violent protracted conflict is thus deemed the result of incompatible interests and/or competition for scarce power resources, especially territory' (Reimann, 2004).
8 In both Manipur and Mizoram, '8 million trees and over 4 million bamboo groves' (Yumnam, 2013) will be destroyed.
9 'Conflict transformation is best described as a complex process of constructively changing relationships, attitudes, behaviours, interests and discourses in violence-prone conflict settings' (Berghof Glossary of Conflict Transformation, 2012).
10 A flagship program of the ruling Indian government, the National Rural Employment Guarantee Scheme (NREGS), aims to generate income and gainful employment for the rural population of the country.
11 'Reconciliation can be thought of as the restoration of a state of peace to the relationship, where the entities are at least not harming each other, and can begin to be trusted not to do so in future, which means that revenge is foregone as an option' (Santa-Barbara, 2007:174).

References

Benedikter, Thomas (2007). *The World's Working Regional Autonomies: An Introduction and Comparative Analysis*. London, New York and Delhi: Anthem Press.

Berghof Glossary of Conflict Transformation (2012). 'Conflict Transformation–Theory, Principles, Actors'. Accessed 13 February 2014: http://www.berghof-conflict foundation.org/images/uploads/berghof_glossary_2012_03_conflict_transformation.pdf.

Bhaumik, Subir (2007). 'Insurgencies in India's Northeast: Conflict, C-option and Change'. East-West Center Washington Working Paper No. 10. Washington: East-West Center.

Boege, Volker (2006). *Traditional Approaches to Conflict Transformation – Potentials and Limits*. Berlin: Berghof Research Center for Constructive Conflict Management.

Caselli, Francesco and Wilbur John Coleman II (2006). 'On the Theory of Ethnic Conflict'. Working Paper 12125. Cambridge, MA: National Bureau of Economic Research.

Chatterjee, Partha (1994). *The Nation and Its Fragments: Colonial and Post-Colonial Histories*. New Delhi: Oxford University Press.

Chaube, S.K. (1999). *Hill Politics in Northeast India*. New Delhi: Orient Longman.

Cohen, Abner (1969). *Custom and Politics in Urban Africa*. Berkeley: University of California Press.

Das, Veena (2007 ed.). *Handbook of Indian Sociology*. India: Oxford University Press.

Geertz, Clifford (1996). 'Primordial Ties'. In Hutchinson, John and Anthony D. Smith (eds.), *Ethnicity*. London: Oxford University Press. pp. 40–5.

Goswami, Uddipana (2010). 'Armed in Northeast India: Special Powers, Act or No Act'. *The Peace and Conflict Review*, 4(2): 25–38.

Guha, Amalendu (2000). *Jamidarkalin Goalpara Jilar Artha-xamajik Abastha: Eti Oitihaxik Dristipat* (in Axamiy) (Socio-economics of Goalpara in the days of the Zamindars). Guwahati: Natun Sahitya Parishad.

Gurr, Ted (2000). *Peoples versus States: Minorities at Risk in the New Century*. Washington, DC: United States Institute of Peace Press.

Hughes McDermott, Melanie (2000). 'Boundaries and Pathways: Indigenous Identity, Ancestral Domain, and Forest Use in Palawan, the Philippines'. Unpublished graduate dissertation, University of California, Berkeley, University Microfilms International.

Hutchinson, John and Anthony D. Smith (1996) (eds.). *Ethnicity*. London: Oxford University Press.

Killick, Nick, V.S. Srikantha and Canan Gündüz (2005). *The Role of Local Business in Peace-Building*. Berlin: Berghof Research Center for Constructive Conflict Management.

Kipgen, Nehginpao (2013). *Huffington Post*, 28 January. Accessed 13 February 2014: http://www.huffingtonpost.com/nehginpao-kipgen/intricacies-of-kuki-and-naga_b_2531115.html.

Murat, Bayar (2009). 'Reconsidering Primordialism: An Alternative Approach to the Study of Ethnicity'. *Ethnic and Racial Studies*, 32(9): 1–20.

Reimann, Cordula (2004). *Assessing the State-of-the-Art in Conflict Management – Reflections from a Theoretical Perspective*. Berlin: Berghof Research Center for Constructive Conflict Management.

Santa-Barbara, Joanna (2007). 'Reconciliation'. In Webel, Charles and Johan Galtung (eds.), *Handbook of Peace and Conflict Studies*. London and New York: Routledge.

Sethi, Nitin (2006). 'Tipaimukh Dam in Manipur Driving a Wedge?' *Down to Earth*, 15 October 2006. Accessed 12 February 2014: http://www.downtoearth.org.in/content/tipaimukh-dam-manipur-driving-wedge.

Six Schedule (2014). 'Six Schedule in Indian Constitution Amendment'. Accessed 13 February 2014: http://lawmin.nic.in/coi/SIXTH-SCHEDULE.pdf.

Tambiah, Stanley J. (1989). 'Ethnic Conflicts in the World Today'. *American Ethnologist*, 16(2): 335–49.

Vanglaini (2011). 'Micro Hydel Inaugurated'. *Vanglaini* (Mizo daily), 8 January, 2011. Accessed 13 February 2014: http://www.vanglaini.org.

Vasquez, John A. and Brandon Valeriano (2009). 'Territory as a Source of Conflict and a Road to Peace'. In Bercovitch, Jacob; Victor Kremenyuk and I. William Zartman (eds.), *Handbook of Conflict Resolution*. London: Sage Publication.

Yumnam, Jiten (2013). 'Tipaimukh Dam Plan and its Uncertainties'. *Imphal Free Press*, 13 October 2013. Accessed 13 February 2014: http://www.ifp.co.in/nws-17592-tipaimukh-dam-plan-its-uncertainties/.

Chapter 3

Self-management and development institutions of Northeast India

J.J. Roy Burman

Introduction

Self-management in the present context is mostly looked at from the viewpoint of either management of the commons or Common Property Resources conceived by the western oriented environmentalists, academicians and Non-governmental Organisation (NGO) personnel. Another breed of academicians thinks of self-management in terms of co-option of the people into the state-sponsored or donor-sponsored programmes and projects and there on participatory development. In the context of Northeast region of India, self-management has been conceived of for quite sometimes and the state in some places has acceded to a large extent the demands of local populace who sought autonomy and self-determination, especially in the hill or tribal societies. With this background the study of functioning of the Sixth Schedule of the Constitution of India gained serious attention. Scholars like Roy Burman (2002), Gassah (2002) and others have examined the functioning of Autonomous District Councils under the purview of the Sixth Schedule in the region in quite detail. While many faulted with its functioning others gave credence to the space offered by the Indian state to different peripheral communities to exercise their autonomy and raise revenue for local self-governance and development. But if one searches in depth, such avenues of self-determination and autonomy are nothing but a façade of state-sponsored or state-regulated autonomy. These, in reality, operate within legal provisions of the state. Political leaders in the Sixth Schedule Areas are now themselves vying for direct contact with the international funding agencies to accrue funds for regional development.

Traditional institutions and self-governance in Northeast India

There is another dimension of self-management that is very profoundly operational in the region which has not gained equal attention. It is the people's institutions involved in self-development, which receive little or no aid from external agencies. These institutions, at times, moderate the entire lifestyle of the communities in the region. Such institutions through the principles of self-determination lead to self-governance as well as self-development. While the government or public mode of development mainly operates through external funding, with no plough-back mechanism, that is the government programmes operate just as fund-disbursing agencies without leading to any local revenue generation through tax, the self-management institutions do not require any external funding. They operate mostly as self-help bodies. In the case of the much talked about the Village Development Board (VDB) institution of Nagaland too, though the local community plays an important role in the selection of beneficiaries, it has hardly any role in the mobilisation of resources. The VDB mainly operates as a fund-disbursing agency and the funds are channelised through the National Bank for Agriculture and Rural Development (NABARD). Quite a substantial amount has been loaned to the VDB by the World Bank. Similarly, the assistances given by the Asian Development Bank are routed through it. Since the people are not responsible for repaying the loans and the Central Government takes responsibility in repaying the loans, the transparency level, too, becomes questionable in the region.

Traditional institutions and development in Mizoram

In Mizoram, though the government has a number of departments promoting welfare activities and construction work, the revenue generation for the state is virtually nil. Therefore, people-oriented programmes become virtually impossible. The most significant incongruity that has emerged is due to the banning of the traditional *jhum* or shifting hill cultivation by the present government in the state. This pushes the poor people virtually to the state of starvation. Horticulture is mainly promoted by the state as the means of production in the rural areas and this does not permit economic liberty as the market is usually controlled by the Marwaris, but during glut, the prices of the produces fall like anything. The industry as it is being promoted is no alternative succour to *jhum*. Naturally Mizoram state on its own is unable to provide any stable welfare institution. Alternatively, the lives in the Mizo-dominated areas of the state are totally influenced by organisations like Young Mizo Association (YMA), Mizo Hmeichhe Insuihkhawm Pawl (MHIP), Zoram Upa Pawl (ZUP) – Old Men's Association and Mizo Zirlai Pawl (MZP) – community-based organisations which surfaced at the time of Mizo upsurge when majority of the villages in the north and central districts of the state was regrouped and placed under the administration of the Army and state bureaucracy and the traditional social institutions that guided the daily lives of the people were completely decimated. Even later, after the formation of Mizoram state in 1985 and the introduction of the District Councils and Village Councils, the situation did not change for the better. The Village Councils are bereft of funds and do not even have meeting places for its members to assemble and take any decision (Prasad, 2002: 217–234).

Presently, every hamlet of all the Mizo settlements has branches of all the three organisations. Membership of these organisations is mandatory and no Mizo can dare to escape them. While the MHIP takes care of issues of the women and children, the YMA mobilises the youth; the ZUP handles the problems of the elderly. The youth of every hamlet congregate at nights for singing and merry making. The young boys and girls join their respective YMA branches also to carry out social work. They are the ones who take charge of repairing the burnt and dilapidated houses of the old and destitute. In case of emergency in a village the YMA members help the people. The YMA is a legacy of the erstwhile Zawlbuk or Bachelor's Dormitory, the institution which was banned by the Christian missionaries in the early phase of proselytiation. YMA is organised and is used as an agency for educating the public in hygiene and cleanliness. The government agencies and the Christian missionaries had made good use of these organisations to improve the standard of living of the Mizos (Lalkima, 1997: 41).

It is true that though the ZUP is not a political organisation, it does take a lot of interest in the political affairs of the state. For instance, it was opposed to the idea of further vivification of the state and did not support the formation of a separate district for the Chakma tribes. ZUP members are also concerned about the efficient functioning of the administration. In 1977, the Mizo Pawls (organisations) gave a call for streamlining the administration and urged the youth to work hard to cover the food deficiency witnessed by the state every year. It also resolved that wages be paid on hourly basis. ZUP also advocated the formation of a Municipality Board for efficient administration of Aizawl town (Ibid.: 56). The MHIP played a crucial role in the formation of

Customary Law Act for the state. Their members were inducted into the state-level committee to assess the status of the women within this Act. The emergence of the new institutions became particularly important in the absence of any revenue-generation mechanism and ploughing back possibilities of the state. The new institutions are further benefited by the altruistic bodies of different churches and the fundamental *mantra* of 'Tlawmgnaihna' embedded in Mizo philosophy. The state, too, has realised the importance of these bodies and is involving them in implementing many of its schemes and programmes.

Traditional institutions and development in Manipur

In Manipur, though the valley, largely inhabited by the Meitei community, is in the throes of violent militancy and rampant extortion activities by the underground. The society, by and large, has been able to retain its sanity and prevent it from miring into a state of anarchy. This could largely happen due to the retention of the principle of clan exogamy. Each of the erstwhile seven principalities of the valley has attained identity as a separate clan. Thus, the valley is to a large extent politically regulated through the principle of kinship and other village-level ethnic bodies. The age grade system and gender bodies like the *meira paibi* (women torch bearers to prevent any antisocial elements) further reinforced the social solidarity at village and *leikai* (local) level. But more than these, the system of *marup* or cooperative groups hinges around different stages of life like birth, marriage, funeral and also gender and age grades. People in a locality come together to form a cooperative, contribute small funds regularly through which they purchase required goods (especially for marriage, dead ceremony or any other special occasions) turn wise. Besides, the intra-village and inter-village level *marups* bridge social bonds within the society. These *marups* based on different enterprises like tin-roof construction of the huts, purchase of motorbikes and refrigerators add on to the cooperative nature of the community. The *marup* differs from the government and other donor-driven self-help groups in that here the entire lives of the people are bridged through self-generated resources, leading to a kind of self-management. It ensures the people to acquire possessions they need most through their own efforts rather than depending on the government subsidies or bank loans. Roy Burman (2002) has also indicated that constituted ethnic formations in the valley, too, have often led to the emergence of self-management bodies. He cited the illustration of Kamlathabi village in Chandel district where five tribes have come down from the adjoining hills and by generating local resources from the land and forests they are running schools, colleges and medical dispensaries. The subjects taught depend on the public demand and resolutions passed by the local management committees. The faculty and the staff also do not have to depend on the government largesse for their wages and salaries.

Similarly, the Loi community (considered to be a Scheduled Caste group) in Manipur valley, residing mainly in four villages close to the foothills, have their parliamentary body to take decisions at the political level. Besides, each of the four settlements has their Village Civic Committees which administer not only the socio-religious responsibilities but also generate revenue through taxes on natural resources and transit tolls. At Sekmai town, about 20 km on the north of Imphal, the state capital of Manipur, an apex level organisation called *Phamnaiba* is in operation. It is a kind of village parliament and constituted of an elder's board. The village headman or Khullappa is the chief functionary of the *Phamnaiba*. All the major village matters are decided by this body. Apart from the *Phamnaiba*, a Scheduled Caste Development Committee has been formed. The Committee encompasses almost all the individuals except for a few Meiteis (general category Meiteis, not Loi or SC category) and Christians. The Committee through almost its own funds has constructed many roads and culverts in the village for public convenience. This apart, has constructed a large Town Committee building mainly by generating its own resources. Sekmai has 21 *leikais* or localities and each of them is managed through a *singlup* or management committee. Every household is mandatorily a member of the *singlup*. The *singlup* members also help in organising weddings and ceremonies in individual households. Weddings involve elaborate arrangements and are quite expensive. Without group support on such occasions people can become paupers. Other than this, the entire Sekmai community is divided into a number of age grades and each of them is assigned certain well-defined responsibilities during the religious functions. Their roles are assigned during village marriages and funerals as well. During the *Khuminpida* ceremony in which married and unmarried daughters offer feast once in a lifetime in order to convey their respect to their elderly fathers and mothers, the *singlup* members take special care to sacrifice pigs and cook them to serve the guests. The *singlups* try to consolidate their unity by organising annual picnics in the precincts of the village. The Lois of Sekmai played a bridge-buffer role between the hill Nagas and the Meiteis of Manipur valley. Their autonomy was a prerequisite for cushioning off conflicts between the two ethnic entities and also bridging the possibility of trade linkages. Interestingly, during the Lai Haraoba festival at Sekmai (also in Meitei) there is a particular ritual in which the presence of persons in traditional Tankhul and Rongmei (categorised as Naga tribes of Manipur) dress is indispensible. Also, the Loi religion and culture share many traditional traits of the hill communities of the state (tribes). The Lois of Sekmai also follows the 'Mangai' system whereby ritual friendship is established between tribes (e.g. Nagas) of adjoining hills. When these hill-men come down for marketing, they usually halt at the houses of the ritual friends. During visits they offer certain gifts to the friends which they have brought from their native villages. The hosts too reciprocate by offering some packed food to their friends for their return journeys. These systems help in maintaining communal amity through people's own efforts rather than depending on state intervention. In the Manipur valley, the *meira paibi* is basically a social reform body which works against alcoholism, drug addiction and antisocial activities. Besides, it also tries to broach peace in the state by negotiating between the underground militants and the security forces. There is a branch of *meira paibi* at Sekmai as well. However, unlike other regions of the valley, the local branch is not opposed to alcohol, since a very large number of village women earn their livelihood by distilling local liquor. Sekmai in this manner even defies the prohibitory order of the State.

The *Paite* is a transborder tribe inhabiting the hills of Manipur, Mizoram and Myanmar. The group largely depends on swidden cultivation, terrace wet rice cultivation, horticulture and timber. Cross-border transactions between the *Paites* of different states are very frequent. The *Paites* of Manipur and Mizoram have an apex level Tribal Council based in the town of Churachandpur in Manipur. The council has its branches at every block and *Paite*-inhabited villages. The youth wing of the council is called *Young Paite Association*. A *Paite* women's organisation similarly exists in every *Paite* domain. While the *Paite* council is mainly involved in political activities, the other body is involved in social welfare by generating their own funds through voluntary labour or cultivation of crops and generating the forests. The Paite Tribal Council (PTC) organised a grand feast of the community after the tribe was assigned a separate ST identity by the Manipur State. During the occasion, Paites from Mizoram, too, flocked in large numbers at Churachandpur. Now, the PTC is actively working in Mizoram so as to accrue a similar ST status in that state. Thus, the council is not merely involved with self-help activities but also at the helm of political affairs and plays a crucial role in political decision-making processes for the Paite community as a whole. The Paite self-management system is further bolstered by the institution of *indongta*. The *indongta* is primarily based on kinship ties. The male head of a family on marriage forms an *indongta* and he is duty bound to assist his younger brothers and sisters during different needs, like he has to help them to construct a house or marrying his children. The father's sister, too, has to assist members of her *indongta* in many different ways (Khamkenthang, 1988). Roy Burman (1988: vii–xix) also writes that the Paites recognise that like beads in a garland the households in a village are tied up with one another by invisible strings of reciprocal obligations and expectations. The obligations and expectations are differentiated in terms of structural relationships with members of wife taking, wife giving and ego's lineage as well as with selected persons of the village who do not belong to any of the categories. Every Paite household has its own *indongta*. At the time when a son or brother separates from his parental home, an *indongta* is instituted for him on the initiative of his father or elder brother as the case may be. It is a formal organisation with recognised functions. The Paites also establish a network of relations through declarations of 'Blood Brother' or 'Zawl'. The Zawls are ritually bound to help each other during any form of distress or even advancing the material quality of life. This enables them to be self-reliant and not be dependent on the state for their well-being.

Traditional institutions and development in Nagaland

The transnational ethnic organisations are to be found in large numbers in the case of Nagas in Manipur, Nagaland, Assam, Arunachal and Myanmar. The Zemis and Zeliangrongs have associations cutting across Manipur, Nagaland and Assam. They are mainly involved in carving out a united Zeliangrong niche or getting affiliated with Greater Nagalim. The Tankhuls of Manipur, Maos from the same state and Nagaland and the Chakhesangs and Angamis mainly supporting the NSCN (IM) have their own Hoho or parliamentary body which helps in functioning of the Naga Federal Government (underground government). The Khaplang faction of NSCN, supported mainly by the Ao, Konyak and Sumi, has a Hoho that has its writ spread even over the Naga areas of Myanmar. Both the factions of NSCN generate huge funds through different forms of taxes and other sources. Almost every Naga village has its own Hoho as well. Besides, in recent years, each village has developed its own development and welfare societies to cater to the needs of different sections of the society. It is a common sight in the newspapers where the village associations announce the names and photos of meritorious students earning laurels in the exams. The self-management systems prevail among most of the Naga tribes almost similarly. The age grade is an ancient tradition of these warrior tribes. The entire society is ranked according to age groups and as one enters into it after attaining puberty; they acquire different ranks till the end of their lives. Different roles are assigned to the different ranks in an integrated manner. Different age grades may accrue funds through collective farming or collective labour or through sundry means. The funds so generated are used for the welfare of the society.

Traditional institutions and development in Arunachal Pradesh

In Arunachal Pradesh, too, the traditional Village Councils presented many elements of self-management. Among the *Khamtis* there exists a self-management institution called *Mokchup*. It is basically premised on the Chiefship pattern. It takes care of all the village-level activities like observance of religious ceremonies, construction of public house or irrigation channel, management of funds raised for a particular purpose and arrangement of voluntary labour for constructional or developmental works, for implementing council's decisions. Its' main function is of course administering justice in the village. Talukdar (2002: 16) concludes about Arunachal Pradesh that there are several types of traditional self-governing institutions among tribes of the state and they fit into different theoretical models. They are normally village-oriented institutions. All these systems invariably include a council of the people or their representatives in one form or the other. The chieftaincies are of milder type and are never autocratic. The Chiefs are often chosen by the people although they come from a particular clan or family. All the types have certain common features in them and function as some sort of village government. The introduction of the statutory panchayat has affected them to some extent but has not replaced them. They still command respect and spontaneous obedience from the people in almost all areas. They have vast experience in managing the affairs of the village. They have successfully carried out activities like making roads, building schools or digging irrigation channels, besides administering justice and maintaining order in the village. A number of unregistered Village Welfare Societies have surfaced in recent years in almost all parts of the state. These groups are generating resources through taxes over natural resources and collection from the vehicles. One such organisation is located at a *Nishi* village in Papum Pare district, about 10 km from Doimukh on way to Ziro. It collects toll tax from every vehicle passing through it. The funds are used to meet the needs of the orphans, widows and the needy people. An annual community feast is a must to regenerate social feelings. These bodies run parallel with the traditional village councils or *Buliangs* involved in self-governance. At *Hong*, an *Apa Tani* village, a self-initiated 'All Forest Preservation Committee' charges royalty on timber, wood posts, coal sand, boulders, gravel and other things collected from the adjoining forests. At *Siro*, a satellite of Hong, All Siro Managing Committee, an unregistered body of which all adult villagers are mandatorily made members, framed a law in the year 2000 whereby people are banned to use poison, chemical and dynamite sticks for fishing. People are also forbidden to enter into the village forests with gun without obtaining permission of the management committee. Anyone violating the rules is liable to be punished and pay fines to the tune of Rs 10,000. For carrying boulders and sand out of the village, the committee charges levy. The committee appoints a person bidding highest for collecting the revenue for a period of two years. The funds are used for the welfare and upkeep of the needy elderly, women and children. Among the *Adi Padam* tribe of Damroh village of Upper Subansiri district, there is a supra-clan organisation (DAME), which was formed at around ten years back. The function of this organisation is to control indiscriminate cutting and felling of timber. It checks stealing of fish and other violations over the use of natural resources by outsiders. The organisation regulates the hunting of games and collective fishing. Regulating the people over the use of natural resources does not prove to be very difficult as such practices are usually confined to clan-owned lands. Worship of a number of sylvan deities like *Aran puja*, *Pucma puja* and *Luiter puja* within the forests further reinforces the informal self-regulation mechanism of the community. Symbolical gestures like prohibition of entering the forests for five days during the *Lune puja* or *Gabbi Holung puja* too metaphorically augment the possibilities of prudent use of nature.

Traditional institutions and development in Assam

In Assam the *Namghars* are present in almost every hamlet of Hindu villages. They are affiliated to different *Satras* or monastic orders. The orders wield significant influence on the daily life of Hindu ethnic Assamese people. The *Namghars* are usually a small covered shed with a *Manikut* or a temple located at one end. It is believed that *Namghar* is a living institution and for over 500 years, its input on Assamese society and culture has been tremendous. It diffused a high degree of enlightenment among the masses of the people. It should be noted that Vaishnavism in Assam is a religion as well as an institution, and even today it exercises a very great influence on the social communal life of the Assamese people. The doors of the *Namghar* are open to all, no matter what caste or gender one belongs to. In the villages, in addition to serving as the common prayer hall, it also serves as the village stage and the meeting place of the village panchayat. It has continued to be the centre of social and religious activities. The *Namghars* are serving till now, as a panchayat hall where the villagers gather, also, to discuss many current problems of the village and community life and political as well as economic and social subjects. The institution helps in unifying the Assamese folk society.[1]

The Churches in the region too are encouraging in many ways for the people to become self-reliant. In some places they are providing the seed money to initiate self-help groups, and in other places the Church members are taking up income-generating programmes through group labour or group farming, piggery and afforestation drives. 'Jirsong Asong' in Diphu or the 'Bosco Reach-Out' in Guwahati in Assam and Shillong in Meghalaya are two of the premiere Catholic groups involved in such endeavours.

Conclusion

The emergence of so many self-management institutions in the region is not surprising considering the fact that strong centres of authority with monarchial power and wealth have never been witnessed in the course of history. Much of the region was ruled through the tenets of anarchic moral orders rather than coercive powers of the state. The rulers were at best ethnic managers controlling the trade and flow of commodities. Sreemanta Sanker Dev, the revered saint from Assam, succeeded in bridging the diverse ethnic entities by propagating a liberal vein of Hinduism. The traditions remain largely unaltered due to a political vacuum created by the Indian nation state which has imposed a colonial legacy of constituency formation whereby the eight states of the region have only twenty-five representatives in the Lok Sabha out of the total of 545 members. The regions having deep roots of linkages with South-East Asia, China and Bangladesh are compelled to evolve their own systems of self-management as a survival strategy. The anarchic states existing in the entire *Zomia* area that includes South-East Asia and Northeast region of India have been recently researched by James Scott and published in his book, *The Art of Not Governing*. The book of course ascribes more importance to the practice of shifting cultivation in the hills and dyadic relations with the wet paddy cultivators in the valleys as the root cause of non-state or quasi-state formations in the region.

The systems of self-management across frontiers can be applicable as in the lines of Sami parliament in the Scandinavian countries. The trans-frontier bodies mentioned here are hooked not only culturally but also politically. Decisions made by the Naga Hoho are equally applicable to the Nagas of both Nagaland and Manipur. The apprehensions shown by certain Assamese scholars like Baruah (2003) towards the Sami model need to be dispelled. Not only the case of Naga Hoho, but also the political decisions of the Paite Tribal Council rule the roost both in Manipur and Mizoram. As it is, the writ of NSCN (K) runs in Nagaland, parts of Arunachal Pradesh as well as the bordering areas of Myanmar. The leader, Khaplang, himself belongs to the Sema tribe which is predominant in the Naga-inhabited areas of Myanmar. These self-management bodies are not confined to just village- or community-level developments. They affect the entire ethnic body right from the grassroots to the top. In material terms, they help the communities meet their felt needs rather than implementing unwanted exotic government schemes and programmes just for the sake of accruing funds. Such external easy funds often lead to the destruction of the social fabric and community bonds that sustain the society. Theoretically, one can probably examine the process through the perspective of 'regulated anarchy' as conceived by Max Weber or Gandhian percept of the minimal state. The anarchism of Bakunin, Proudhon and Kropotkin in the Western mould based on colonial industrial economy would not perhaps be so appropriate. The region does not sustain by colonising others, and is rather a subject of internal colonisation. The self-management institutions here may be conceived of as symbols of resistance against the state and global hegemonic onslaught. Not much work has been done in this regard, particularly, in terms of the developments in the third-world countries or postcolonial countries. Even the postmodernists from the west have no answer to such situations. The bonds depicted here are mainly in terms of intra-societal dynamics. There are also possibilities of inter-village, inter-ethnic and even interregional institutions that provide a sense of stability. But I have had no occasion for examining them as yet.

Note

1 Retrieved from http://www.atributetosankaradeva.org/namghar.htm.

References

Baruah, S. (2003). 'Confronting Constructionism: Ending India's Naga War'. *Journal of Peace Research*, 40(3): 321–38.

Gassah, L.S. (2002). 'Traditional Self-Governing Institutions among Hill Population Groups of Meghalaya', In A. Goswami (ed.), *Traditional Self-Governing Institutions among the Hill Tribes of North-East India*. New Delhi: Akansha Publishing House.

Khamkenthang, H. (1988). *The Paite: A Transborder Tribe*. New Delhi: Mittal Publishers.

Lalkima, C. (1997). *Social Welfare Administration in a Tribal State: A Case Study of Mizoram*. Guwahati: Spectrum Publications.

Prasad, R.N. (2002). Restructuring and reform of rural administrative institutions in Mizoram: A study in the new panchayat raj system, In G. Palanithurai (ed.), Dynamics of new panchayati raj system in India, Vol. II. New Delhi: Concept Publishing Company, pp. 217–234.

Roy Burman, B.K. (1988). Forward, In H. Khamkenthang (ed.), *The Paite: A trans-border tribe of India and Burma*. New Delhi: Mittal Publications, pp. vii–xix.

Roy Burman, B.K. (2002). 'Traditional Self-Governing Institutions among the Hill Tribal Population Groups of North-East India', In A. Goswami (ed.), *Traditional Self-Governing Institutions among the Hill Tribes of North-East India*. New Delhi: Akansha Publishing House.

Scott, J. (2009). *The Art of Not Being Governed: An Anarchist History of Upland South-East Asia*. Yale: Yale University Press.

Talukdar, A.C. (2002). 'Traditional Self-Governing Institutions among the Tribes of Arunachal Pradesh'. In A. Goswami (ed.), *Traditional Self-Governing Institutions among the Hill Tribes of North-East India*. New Delhi: Akansha Publishing House.

Chapter 4

Vicious circle of insurgency and underdevelopment in Northeast India

P.R. Bhattacharjee and Purusottam Nayak

The insurgent groups

Northeastern region (NER) of India comprising eight states namely Arunachal Pradesh, Assam, Manipur, Meghalaya, Mizoram, Nagaland, Sikkim and Tripura has earned the dubious distinction for persistent underdevelopment and growing insurgency. The fire of insurgency has engulfed the region in such a way that there seems to be existence of a parallel authority of the insurgents in many places as rampant abductions, extortions and killings go on unabated. Consequently, normal life is often paralysed and all initiatives including the socio-economic ones are increasingly crippled as an air of fear and uncertainty pervades the region. This chapter makes an attempt to explore the two-way linkage between insurgency and underdevelopment in Northeast India.

Nagaland has been the epicentre of insurgency in the region. The Naga leader, A.Z. Phizo, raised the banner of revolt (under the banner of Naga National Council) at the very dawn of Indian independence, claiming that Nagaland had never been a part of India. Although the subnational state of Nagaland was created in 1963 in order to fulfil the political aspiration of the Nagas, the flame of Naga insurgency could never be doused effectively and now it affects Manipur, Assam, Arunachal Pradesh and Nagaland as the Naga insurgent outfits aim at political union and independence of all the territories claimed to be Naga-dominated areas. Also, Naga outfits are providing help and training to the insurgents in other states of the region as well. Lately, the National Socialist Council of Nagaland (NSCN), formed in 1980 (now split into two factions), is the most formidable insurgent outfit in the region.

In Assam, the insurgency has grown out of mass movement over the foreigners' issue which started in 1979. The United Liberation Front of Assam (ULFA), which has been committing terrorist acts with their avowed objectives of forming independent Assam, has created a serious internal security hazard. The Bodo (one of the leading tribal communities) of Assam has also taken up arms. Similarly, in the North Cachar district, the Dima Halam Daogah (DHD) is engaged in insurgency activities. The declared political ambition of the Bodos is the formation of a separate state under the Indian Union and further to attain independence, while the objective of DHD is not explicitly made known. Thus, Assam faces a very complicated problem of insurgency.

Manipur is plagued by triple problems. The valley faces the insurgency of the Meitei militants for sovereignty while the hill areas are affected by depredations of the Naga militants on the one hand and Kukis on the other for their separate statehoods. The more prominent militant groups operating in the state are the People's Liberation Army (PLA), United National Liberation Front (UNLF), People's Revolutionary Party of Kangleipak (PREPAK), Kangleipak Communist Party (KCP), Kanglei Yawol Kanna Lup (KYKL), Kuki National Organisation (KNO) and NSCN (Singh, 2000).

In Tripura, the tribal and non-tribal divide has generated dissension from the very dawn of the state's accession to the Indian Union in 1949. Although socio-economic development has traditionally been associated with immigration, the massive influx of the non-tribal refugees from East Pakistan (now Bangladesh) in the wake of the division of India in 1947 reduced the tribal people into a minority, constituting less than one-third of the population of the state. As most of the immigrants settled in rural areas, the pressure of population on land has been tremendous. The sense of being progressively marginalised gave rise to insurgency. In the 1980s, the Tripura National Volunteers (TNV) was a formidable militant group spreading hatred against the non-tribal and it was mainly responsible for the riots that took place in June 1980. At present, there are about 20 militant groups, the two prominent ones being the National Liberation Front of Tripura (NLFT) and the All Tripura Tiger Force (ATTF). In the recent past, the non-tribal groups are also being involved in anti-tribal violent activities (Ganguly, 1999).

Mizoram experienced rebellion of the Mizos under the leadership of Laldenga under the banner of Mizo National Front (MNF). But after the Mizo Accord of 1986, there has been no major wave of insurgency. The issue of insurgency movement has almost died down. Nonetheless, inter-tribe conflicts and suspicion against the non-tribal are not altogether absent. Hmar tribe demands separate state and Brus have not forgotten the atrocities of Mizos. The same more or less holds true in Sikkim as well.

Meghalaya had been free from insurgent activities for long. However, for the last few years Hynniewtrep National Liberation Council (HNLC) has been organising disruptive activities in the state in spite of the fact that they do not have mass support. The Garo National Council (GNC) and the Garo Students' Union (GSU) are demanding a separate state for the Garos on linguistic lines. Arunachal Pradesh, regarded as an island of peace in the region, is not totally free from trouble. The local people refuse to allow the Chakmas to be absorbed in the state's population. There has been a rising trend of ethnic separatism. Demand for two autonomous district councils – Mon and Patkai – has rocked the state, and the same was opposed by major tribes.

Insurgency and economy

As security is the primary infrastructure of economic activities and social stability and certainty about future are the prerequisite of investment, the persistent insurgency atmosphere has been the most important contributor to economic stagnation of the region. It is to be noted that the gateway to the region is the Siliguri area in North Bengal and all flows to and from the region on the surface routes have to pass through this area and the Brahmaputra Valley of Assam, Guwahati being the grand nodal point. Therefore, any disturbance in the Brahmaputra Valley and/or its adjoining hills brings the activities in the whole of the region to a standstill. Also, there is important spillover effect of insurgency in one state on the contiguous states. Therefore, the problem has to be viewed and tackled in an integrated manner considering its regional external effects, uniformity in its basic nature and also the linkage between the insurgent outfits of different states.

So far as the economy of the region is concerned, the first casualty of insurgency has been its already weak infrastructure, especially its transport system. The subversive activities of the insurgents damage rail tracks, cause accidents leading to loss of life and property, create terror among the travellers and throw the entire system out of gear. Similarly, vehicles in the state and national highways are often attacked, passengers and transport workers are killed or wounded and sometimes abducted for ransom; and goods are looted. As the region suffers from geographical isolation and faces ravages of nature in the form of floods (in low-lying area of Assam) and landslides during monsoon (hill tracks), the insurgents' attack on the transport artery represents the last straw on the camel's back.

The second important target of the insurgents is the resource-based industries like petroleum and tea which form the core of the modern organised sector in the region, especially in Assam. As the articulated economic grievance hovers around the idea of so-called regional colonialism based on the alleged drain of rich resources of the region, any violent political movement makes petroleum and tea as its target. Oil pipelines are often blown up by the insurgents, tea gardens are targeted for extortion and sometimes tea garden executives are abducted. Tea gardens constitute the soft targets of the insurgents as these are in the vicinity of forests and away from the populous localities in Assam. It is easy to understand that disrupting industrial activities centering petroleum and tea are bound to block the wheel of progress in the region. The attack of the insurgents on tea and petroleum industries is bound to convey negative signal to the prospective investors. The potential of using gas reserve of the region will also be seriously hampered because of insurgency movements. Albeit there is fierce competition between the states for attracting domestic and foreign investments in the post-liberalisation period, the region is lagging behind in the country and will face further hurdles in the economic development process.

Third, in the long-term point of view, another victim of insurgency is the environment and natural forest resources. On the one hand, insurgents damage forests by taking shelter there and on the other, anti-insurgency operations also lead to denudation of forests. This not only means that conservation activities and other forestry operations are hampered thus resulting in the loss of valuable natural resources but also a grave threat is posed to the fragile ecology of the region. For instance, in the 1990s, roadside tree plantations have been extensively felled in the highways of Imphal in the pretext that insurgent groups take advantage of the tree for attaching security forces.

Fourthly, insurgency has created serious problems for development of the interior areas. As is well known, the region is predominantly ruralised, characterised in many parts by hilly terrain and sparsely populated. It is extremely difficult to build up rural infrastructure like roads and communication links, power grid and irrigation arrangements. It is also equally difficult to build up and administer schools, hospitals, agricultural extension centres and any other thing in such a condition. In other words, rural–urban economic interaction in the hilly and interior areas of the region has to cross a number of hurdles. The insurgency has aggravated the problem to such an extent that development workers are utterly discouraged from going to the hilly and rural areas as they face constant extortions and threats of abduction or death. Consequently, insurgency is pushing the backward areas of the region to the darkness of greater underdevelopment and is acting as a retarding force rendering disservice to rural poor especially the indigenous people.

In reality, economic development involves structural changes and organisation of new economic activities with new techniques of production. Participation of development organisers, experts and even workers from outside the traditional social boundaries becomes indispensable in such circumstances. The insurgents by breeding a cult of hatred against the supposed or real outsiders are blocking all inflows of resources, ideas, expertise and initiative to the societies of the region. This is bound to tell upon the future of the region. It is a contradiction to grumble about the state of underdevelopment on the one hand and target the agents of development on the other in the name of protection of the interest of indigenous people. Why is this contradiction generated in the behaviour pattern of the insurgents? Surely, the insurgents cannot simply be regarded as disruptionists as they undoubtedly command some degree of popular support directly or indirectly as they get food, shelter and other assistances in their societies and as they derive inspiration from the autonomy movement in the region. In fact, the insurgents may be regarded to represent the violent stream of the ethnic separatist movement having roots in the socio-economic-political grievances. So, their behaviour pattern has to be analysed with the help of social theories. This is attempted in the following section.

Insurgency and social theory

The origin and persistent existence of ethnic separatism as expressed in the form of insurgency may be explained in terms of Amartya Sen's concept of *cooperative conflicts*. According to Dréze and Sen (1999: 11), "In the social relations that inter-alias determines the entitlements enjoyed by different people, there tend to be a coexistence of conflict and congruence of interests. There are in most situations, clear advantages to be gained by different people through co-operation with each other and yet there are also elements of conflicts reflecting the partly divergent interests of the same people. Co-operative conflicts refer to this co-existence of congruence and conflict of interests providing grounds for co-operation as well as for disputes and battles." Co-operative conflicts may be illustrated from many different fields of social relations. An illustration provided by Dréze and Sen (p. 12) is as follows – "Consider the relation between workers and industrialists in a particular industry. If production is disrupted, both the industrialists and the workers may lose, so that it is in the interest of both to cooperate with each other in the process of production. But the division of benefits obtained from production may also involve an extensive tussle between the industrialists and the workers. Another illustration of the working of the theory of cooperative conflict is provided by the relation between different members of a family where cooperation is essential for living together and yet conflict arises in sharing of benefits."

The theory of cooperative conflict is suitable for explaining the insurgency and social dissensions in the region. On the face of it, insurgency is organised against the Central Government, and the states are targeted because of their link with dependence on the centre. The non-indigenous elements of population who are considered as *outsiders* are also made the targets of attack. But, the main grievance arises out of the iniquitous character of the development process of the policymakers (government machineries) which in spite of some measures of welfares for the indigenous people here and there has not been able to tackle their basic problems effectively. On the one hand, the centralised character of planning has deprived the indigenous people in determining the nature and contents of development (thus frustrating their aspiration for autonomy), and on the other, the predominance of the tertiary sector led by the government administration and creation of scattered and a few resource-based industrial islands have left little scope for participation of the indigenous people in the development process and have seriously restricted their entitlements. Therefore, they do not find much gain in cooperating with the Government and the so-called outsiders in promoting development. Neither political stability nor economic development in the present milieu seems to have relevance to their interest as these aims at the development of the geographical areas constituting the region without ensuring significant uplift of the poorest of the poor who constitute the bulk of the indigenous population.

Although the cooperation between different sections of the people and between the people and the Government could lead to development, the political conscious among the indigenous people have chosen the path of non-cooperation and violent disruption. What appears to be revolt against the Central Government for attaining either sovereignty or enlarged autonomy is actually a violent expression of grievance against a political and economic structure in which the indigenous people have little share in policy making and from which they do not benefit to the extent as would have satisfied their expectation.

The turmoil is also a result of lack of cooperation between different sections of population. The broad pattern of social dissension takes the form of indigenous versus the non-indigenous, tribal versus non-tribal, autochthones versus immigrants. But even within the different sections of autochthones, areas of conflict exist as all of them compete for scarce resources like cultivable land, use of forests, lakes and streamlets for economic purposes in a relatively stationary economic perspective. While, in a broad sense, the problem of insurgency emanates from underdevelopment, it is the peculiar features of underdevelopment and development in the region that are crucially important for understanding the causes of insurgency. It is the distribution of burden of underdevelopment and costs and benefits of development among the different sections of the population that are causing dissensions and consequence revolt.

What has been stated in the foregoing paragraphs may well be illustrated with the help of experiences of different states in the region. Let us take the case of Tripura (see Ganguly, 1983; Bhattacharjee, 1993). Economic development of Tripura has been historically associated with immigration from the areas which now constitute Bangladesh. After the partition, refugee rehabilitation made the tribal, the autochthones minority in the state. Much of the political grievances of the tribal and the consequent tribal insurgency are ascribed to this factor. But, the present state of affairs could have been definitely avoided or at least mitigated if in the post-independence period the tribal economy of the state was effectively linked with the relatively dynamic segment of economic structure of the state.

The tribal people found that as economic development gathered momentum, they were losing their traditional rights on forests and other natural resources. They also found that they would have extremely limited scope for wet rice farming if they opted for it by giving up shifting cultivation as arable land had already been scarce. They had little role to play in tea plantation industry as it had its own immigrant labour force. The tertiary sector led by public administration offered some scope to the educated tribal but their participation in this sector was limited by tardy progress in education and attainment of technical and entrepreneurial skill. The potential resource-based industries using natural gas also did not mean much for them. Naturally, they could pin their hope only on the primary sector consisting of agriculture, horticulture, forestry, animal husbandry, fishery and so on. Rubber plantation undertaken on family basis could be a growth booster for the tribal society. But, the relatively long gestation period, the skill gap, distrust of the market (generated by the past bitter experience of exploitation) and the urge for producing own food prevented the tribal people from taking advantage of it in a meaningful scale. Consequently, the tribal were left to the *jhumia-cum-peasant* or *pure jhumia* status (*Jhumia* refers to shifting cultivator). As they lagged behind, the anger accumulated finally bursting into insurgency. Assam has been a classic example of economic development which was deeply influenced by exogenous factors. Development of Assam has been accompanied with massive immigration caused by both economic and political factors (Goswami, 1988). Assam's demography has a truly plural character consisting of Assamese-speaking Hindus and Muslims, Bengali-speaking Hindus and Muslims, indigenous plains tribes, indigenous hill tribes, Nepalese settlers, population historically linked to tea plantation, businessmen, public servants and others coming from other states of India.

Of these groups, the Assamese-speaking Hindus and Muslims with a sizeable middle class have been the dominant group politically. The anti-foreigners agitation mainly aimed against the immigrants from East Pakistan (Bangladesh) was led by people from this group. The movement was apparently organised to protect the sociocultural, economic and political interest of the indigenous population. But its economic roots lay in the economic stagnation resulting in the fierce competition among the middle-class people belonging to different linguistic groups (especially the Assamese and Bengalese) for government jobs on the one hand and increasing pressure of population in the state in general and its valleys in particular. Later on, the ULFA was born with its avowed objective of secession from the Indian Union. Gradually, the wrath of extremists also targeted the businessmen hailing from other states of India and the consequence is a threat to the very future of economic development of Assam.

It is to be noted that the plains and hill tribal people of Assam do not fully feel themselves identified with the movement led by the Assamese-speaking middle class. It is because of this fact that they have separatist movements of their own like the Bodo insurgency and political movements in the Karbi Anglong and North Cachar Hills districts. The different segments of the population thus have their sectarian interests and their co-operative conflict ultimately expresses itself as a revolt against the local and central political establishments.

In the case of Nagaland and Manipur, the problem of sociocultural plurality in the population is not as intense as in Tripura and Assam. But the dualism involving the autochthones and the outsiders has gradually emerged as administrative infrastructure expanded and transport, communication and power development took place. The suspicion about the outsider and the peculiar coexistence of modernity and tradition (the former increasing consciousness about rights and the later forming walls of conservatism) made the indigenous people non-cooperative with the central establishment.

The way ahead

It goes without saying that it is not easy to break the vicious circle of insurgency and underdevelopment existing in the region. It can be nobody's claim that the problems can be tackled solely by administrative, political or economic measures adopted in isolation. It requires not only a holistic approach but also a great deal of patience and understanding of the specific problems of different sociocultural-ethnic groups residing in the different parts of the region.

As we have already emphasized, socio-economic pluralism and inequality in the participation of the development process generating cooperative conflict lie at the root of the insurgency in thc region. Therefore, the distinction between economic development of the geographical areas constituting the region and that of its backward sections must be borne in mind by all those who seek solution to the problem. The conflict between equity and efficiency is very sharp in the region. If equity especially inter-group equity (group referring to a distinct sociocultural-ethnic identity) cannot be ensured, development effort will be thwarted by social revolt. Therefore, development projects must be such that they are able to receive social approval at the grass-root level, given the diversities of population. This means that it must ensure the participation of the masses of all groups in the development process with reasonable equity in the distribution of costs and benefits.

If we agree on the basic feature of development policy noted earlier, it follows that modernisation of the primary sector and a vigorous programme of rural development should be nucleus of all development efforts. Although the region is industrially backward, stress on conventional pattern of industrialisation will do more harm than good as this would largely bypass the indigenous masses at least in the short run. Instead, rural industries must be rejuvenated. If agriculture, animal husbandry, forestry and other industries are organised on the modern lines there is scope for development of a number of processing industries in the region.

It may, however, be remembered that growth of a modern middle class and change in taste and preferences of even the rural people may stand in the way of reviving some of the traditional industries like the tribal loin looms and handlooms. Their place has to be taken by processing industries connected with agriculture, animal husbandry, horticulture and forestry and also service industries related to the rural infrastructure like irrigation, rural electrification, transport and communication and marketing. Family plantation units in rubber and tea backed by processing units at the corporate level will increase the spread effect of these industries, especially in Assam and Tripura.

Another point to be stressed is that development must be ecologically sustainable and must not be eco-degrading. Even schemes for infrastructural development like construction of roads and railways and power projects must be so drafted as to involve minimal adverse effects on the ecology. Besides, special attention must be paid while going for mining and cement industries. It must be remembered that in the region, the masses are overwhelmingly dependent on forests, rivers and other natural resources for their daily requirements. Even plans for environment protection must be chalked out with the consent of the people and keeping their interest in view (Ganguly, 1996). Although statehood under the Indian federation cannot be granted to each subnational group or each sociocultural-ethnic entity, their distinct identities must be respected. The best way to ensure decision-making power to different groups is to decentralise the political and economic administration. Federal principle should be followed not only in the centre–state relation but also in the relation between the State Governments and local-level political administrative units.

In the region, the Panchayati Raj institutions, Autonomous District Councils, and Sub-State Regional Development Councils have not been able to achieve the objective of decentralisation in decision-making process as they have lacked real power and modern outlook and efficiency (Ray, 1999). Serious thought must be given to the task of developing a decentralised power structure not only in the region but also in the whole of India if separatism is to be fought effectively. For the region, it is indispensable that the fear of the indigenous masses regarding loss of freedom and identity must be removed.

Last but not the least, it must be driven home to the people of the region that they can never attain real political power and development if they depend on the outside funding in all important schemes as at present. Even while accusing the centre for neglecting the region one cannot deny the fact that most of the State Governments in the region are financially non-viable and they depend largely on the largesse of the Central Government for both plan and non-plan expenditure as evident from their annual budgets. In this context, the way to achieve greater financial and political autonomy and to avoid non-priority expenditures or wastage is the local resource generation for local development works. Of course, the resource-transfer from the centre can never be totally dispensed with. But if Central resource transfer is matched by significant amount of resource generation at local levels, it will not only reduce inequality and wastage in the region but will also give the people the real taste of political power.

References

Bhattacharjee, P.R. (1993). *Economic Transition in Tripura*. New Delhi: Vikash Publishing House Pvt. Ltd.

Datta Ray, B. (1999). 'Autonomous District Councils and the Strategy of Development in North-East India'. In A. Bannerjee and B. Kar (Eds.), *Economic Planning and Development of North-Eastern States*. New Delhi: Kanishka Publishers and Distributors.

Dréze, Jean, and Sen, Amartya. (1999). *Hunger and Public Action*. Delhi: Oxford University Press.

Ganguly, J.B. (1983). *Benign Hills*. Agartala: The Tripura Darpan Prakasani.

——— (1996). *Sustainable Human Development in the North Eastern Region*. Shillong: ICSSR, North Eastern Regional Centre.

——— (1999). *Peace and Development in Tripura: Problems and Prospects*. Guwahati: Nagaland Gandhi Ashram.

Goswami, P.C. (1988). *The Economic Development of Assam*. New Delhi: Kalyani Publishers.

Singh, P. (2000). 'North-East: The Frontier in Ferment'. *Dialogue*, 1(4), April–June, New Delhi.

Chapter 5

Politics of peace accords in Northeast India

Komol Singha and M. Amarjeet Singh

Introduction

India's Northeast region (NER hereafter) consists of eight States,[1] is surrounded by five countries[2] and connected to mainland India through a narrow land corridor at Siliguri of West Bengal. It accounts for about 8.06 per cent of the country's total landmass and about 4 per cent of the population. The NER is also one of the most underdeveloped parts of the country, inhabited by a large number of ethnic groups. The region is confronted with a large number of challenges simultaneously – armed conflict, ethnic conflict, poverty, inequality, etc. According to Heimerdinger and Chonzom (2012), the discontentment and assertion of ethnic identities are that a large part of the region had remained as a loose 'frontier area' and as a result, 'never came in touch with the principle of a central administration before'. It was after many years the region was brought together under a unified administration by colonial rule. This was further convoluted by the controversial integration of ethnic minority groups into newly independent India and an opaque reorganisation of States where their cultural specificities were ignored (Singha, 2012).

Since independence in 1947, the country's Central Government has been trying to tackle several armed conflicts by means of economic development, political reconciliation and use of military force. Use of significant military force under extraordinary legislations such as the Armed Forces (Special Powers) Act of 1958 has been in place for the last several decades. Unfortunately, instead of resolving the problem, the use of armed force has led to its escalation (Heimerdinger and Chonzom, 2012). The conflict disrupts normal life and administrative activities, and incidences of violence have increased significantly in the last few decades. The armed groups have created parastatal organisations in the areas where they operate and set a certain rules and regulations to protect the means and the resources necessary for their existence. In short, some armed groups have not only monitored the institutions of the State but also administered their own social welfare services. They have set up a parallel government structure in different parts of the region, and have set up their own economic policy and their own court system which provides justice expeditiously.

In addition to the adoption of military force to suppress the rebellion, the Central Government has also promoted political settlement and dialogue which led to the signing of peace accords with different armed groups and other ethnic leaders in the past. But, most of them have failed, or could not claim to be broadly successful. Based on five peace accords, this chapter seeks to evaluate their impact to the specific conflicts. The five peace accords are as follows: 16-Point Agreement (1960), Shillong Accord (1975), Mizo Accord (1986), First Bodo Accord (1993) and Second Bodo Accord (2003). These peace accords have been deliberately selected because of their significance and relevance in the course of conflicts in three States of the region, namely Nagaland, Mizoram and Assam.

Interface between armed conflict and peace accord

An armed conflict is defined as 'a contested incompatibility that concerns government and/or territory where the use of armed force between two parties, of which at least one is the government of a state, results in at least 25 battle related deaths in one calendar year'.[3] It is also a situation in which the parties use conflict behaviour against each other to attain incompatible goals or to express their hostility (Bartos and Wehr, 2002). While the government is generally reluctant to recognise the armed groups' legitimacy as a bargaining partner, some conflicts do reach a point where the two sides open peaceful negotiations. The prospect for negotiation becomes stronger for the armed groups if they survive a certain period of fighting. If the government fails to repress them early, they may be forced to negotiate under less favourable circumstances later in the conflict (Bapat, 2005). As a measure to end conflict, the negotiated settlements since 1990s have been an increasingly preferred method of conflict resolution in different parts of the world. As a result, there has been a proliferation of peace accords, successfully ending many violent conflicts (Bell and O'Rourke, 2007). Negotiation to end armed conflict is a dynamic bargaining process (Höglund, 2008) which is both critical and highly sensitive (St. John, 2008). It involves bargaining between the parties to the conflict where both sides have power over each other (Zartman, 2008).

While conceptualising conflict resolution, the 'Ripeness Theory' of Zartman (2001) is identified as one of the most accepted theories of peace agreement (Tiernay, 2012). It is based on the notion of Mutually Hurting Stalemate, wherein the parties find themselves locked in a conflict from which they cannot escalate to victory and the current deadlock is painful to both of them. As such both parties seek an alternative policy or way out (Zartman, 2001). Henceforth, the parties enter peace negotiations or accords. Negotiation constitutes 'dialogues over time between representative of contesting forces with or without an intermediary, aimed at securing an end to hostility over issues that transcended a strictly military nature' (Arnson, 1999). Negotiation is likely to take place and is fruitful if both parties have some optimism of finding a mutually acceptable settlement, and the forces that encourage negotiation must outweigh the structural changes that keep the conflict escalating (Pruitt and Kim, 2004).

The signing of peace accord usually implies an acceptance of both parties to terminate armed conflict. Thus, the clearest method of ending an armed conflict is through a peace accord (Derek, 2003). Hence, peace accord is the 'formal agreement between warring parties, which addresses the disputed incompatibility, either by settling all or part of it, or by clearly outlining a process for how they plan to regulate the incompatibility'.[4] It can be divided into full, partial and peace process accords. Full accords are those where all warring parties make an accord to settle the incompatibility and where there is no continued fighting. Partial accords are those concluded between some of the parties, but not all. The incompatibility is regulated between the parties concluding the accord. Sometimes fighting will cease, but on other occasions a party not included will continue the conflict. Finally, the peace process accords do not settle the incompatibility, but instead outline a process whereby the issue will be settled (Harbom et al., 2006).

Peace accords, including in Northeast India, do not produce many of the expected benefits. Peace processes always create spoilers because it is rare in armed conflict for all factions to see peace as beneficial. Even if they do, they rarely do so simultaneously. A negotiated peace often has losers, either leaders or factions who could not achieve their initial aims. Nor can every war find a compromise solution that addresses the demands of all the factions (Stedman, 1997). Similarly, in Northeast India, an inclusive peace accord might be effective. As such, signing a peace accord must include all the stakeholders, and the leaders must represent the members of warring groups. If major leaders are not included, obfuscated peace talks can bring broad resentment and the number of groups with competing goals will emerge (Cline, 2006; Ravi, 2012).

Perhaps, the most significant debate in peace studies relates to the definition of 'peace'. It is a 'word of so many meanings that one hesitates to use it for fear of being misunderstood' (Boulding, 1978). It is an 'umbrella concept', a general expression of human desires, of that which is good, that which is ultimately to be pursued (Galtung, 1967). In this context, Gunnar Johnson made the following statement, 'that the image of peace in peace studies is blurred as a result of conflicting definitions held by researchers. . . . The conceptual chaos stems in part from a general tendency to focus on violence and war or other global issues rather than on peace per se and in part from the indeterminate nature of the term itself. Consequently, researchers are currently studying different problems and pursuing divergent goals, all under the banner of "peace"' (Johnson, 1976).

Nonetheless, peace is largely seen as the absence of violence. Johan Galtung introduced and popularised the meaning of 'negative peace' and 'positive peace'. Since then peace has largely been defined either in a negative or a positive sense. Negative peace is 'an idea of peace as the absence of organised collective violence, in other words violence between major human groups; particularly nations, but also between classes and between racial and ethnic groups because of the magnitude internal wars can have.' Thus, negative peace is the absence of violence. In contrast, positive peace is 'a synonym for all other good things in the world community, particularly cooperation and integration between human groups.' Thus, it refers to the absence of structural violence (Galtung, 1967). For Kenneth Boulding, peace is not merely the absence of war but also absence of turmoil, tension and conflict, and the positive peace is 'a condition of good management, orderly resolution of conflict, harmony associated with mature relationships, gentleness, and love' (Boulding, 1978: 3). Emmanuel Adler has, however, described positive peace as belonging to idealists who portray peace as some kind of utopia 'incorporating the improvement of politics and human nature, social justice, morality, international organization and law, and human progress' (Adler, 1998). Thus, 'analytically this division between negative and positive peace seems to be a muddle. However, it can be concluded that negative peace denotes the absence of war and that positive peace denotes something further. But exactly what these "further" things actually would be, is hard to comprehend' (Albert and Carlsson, 2009).

Insights from five peace accords

16-Point Agreement with the Nagas, 1960

The armed conflict in Nagaland (then Naga Hills district of Assam) started soon after India's independence in 1947. The Nagas claimed that their ancestral homeland was an 'independent country' in the pre-colonial period and hence wanted to regain their pre-British status. The Naga Club, an interest group, took the first ever initiative to bring the Naga people together under a single administrative unit. In June 1947, an agreement, known as the 9-Point Agreement, was concluded between the Naga National Council (NNC) (the successor of Naga Club) and the then Governor of Assam to install an interim administrative arrangement for the Naga Hills district. Ironically, it did not materialise as the terms of the agreement were thereafter contested by both sides. Initially, the Central Government was willing to grant limited political and economic autonomy under the Sixth Schedule[5] of the Constitution of India. The offer was not only rejected, but also led to the commencement of the first armed rebellion against India.

In the meantime, a group of intellectuals of the Naga community who wanted a negotiated settlement with India within the country's constitutional framework formed the Naga People's Convention (NPC). Initially, it insisted for the constitution of a single administrative unit comprising Naga Hills district of Assam and Tuensang Frontier Division of North East Frontier Agency. The government accepted the demand and hence an administrative unit of Naga Hills-Tuensang Area was constituted in 1957 comprising Kohima, Mokokchung and Tuensang districts. Consequently, the NPC put forward an additional demand to elevate Naga Hills-Tuensang Areas into a State (Inoue, 2005). Once again, the demand was accepted by the government which came to be known as the 16-Point Agreement of 1960 and hence the State of Nagaland came into being. However, a faction of the NNC vowed to continue to fight for an independent homeland and assumed no obligation to accept the India-imposed Nagaland.[6]

Shillong Accord with the Nagas, 1975

In 1975, six Naga leaders[7] calling themselves the 'Representatives of Underground Organisations' signed an agreement with the Governor of Nagaland, at Shillong, Meghalaya's capital, popularly called the Shillong Accord. They acted at the behest of a faction of the NNC. Accordingly, they agreed to abide by the Constitution of India and surrendered arms.

It was also agreed that they should get reasonable time to formulate other contentious issues for discussion for final settlement. As expected, it was rejected by another faction of the NNC for what they called 'selling the Naga nation'. They said that no member of theirs were involved in signing this accord (Shilling Accord).

Mizo Accord, 1986

The two-decade-old armed conflict in Mizo (Lushai) hills district of Assam came to an end in 1987. The Mizo Union, a political party in Mizo hills, was in favour of integration with India, while the United Mizo Freedom Organisation, another political party, preferred to join Burma. However, when India attained independence, the Mizo hills remained as a district of Assam. When the demand for the reorganisation of States on linguistic basis gained momentum in the 1950s, in different parts of the country, the Mizo Union demanded a separate State comprising Mizo-inhabited areas of Mizo hills district. The States Reorganisation Commission[8] turned down their demand. Based on the Commission's recommendations, 14 States and six union territories were constituted in India in 1956. Accordingly, Northeast India constituted the State of Assam and Union Territories of Manipur and Tripura. Sikkim was not included in this context as it was clubbed to NER fabric in 2003.

The Mizos protested when the Assam Government declared Assamese as the official language for official communication in the State in 1960. Around this time, a severe famine, known as *Mautam*, occurred in the Mizo hills due to the exponential increase in rat population during bamboo flowering season. This damaged crops on a large scale. Relief works carried out by the government agencies was inadequate causing widespread discontent among the Mizo people. Several voluntary organisations came forward to help the famine affected people. Among them, the Mizo National Famine Front did exceptionally well in providing food to the affected people. Consequently, it dropped the word 'Famine' and became a political party in 1961, seeking an independent homeland of the Mizos. The party later transformed into an armed group, the Mizo National Front (MNF). On 28 February 1966, they carried out simultaneous attacks on various places of the district. Subsequently, the group declared independence from India on 1 March 1966. As a result, the Central Government deployed armed forces to control the situation. The district was declared 'disturbed area' under the Assam Disturbed Areas Act of 1955 and eventually led to the imposition of the

Armed Forces (Special Powers) Act of 1958. Consequently, as a measure to contain conflict, the government initiated negotiation with the MNF. Around this time, the Mizo District Council, an elected legislative body of the district, revived the demand for a separate State. The Central Government also partially accepted their demand by elevating the district into a Union Territory[9] of Mizoram in 1972 and also promised to upgrade Mizoram into a State. After the signing of the peace accord between the MNF and the Central Government in June 1986, Mizoram became a state in 1987. This also led to the transformation of the MNF into a political party and the armed conflict came to an end. Mizoram is now considered peaceful (Ray, 1993).

First Bodo Accord, 1993

In the late 1960s, the Bodos, one of the largest ethnic groups in Assam, began their agitation for inclusion of Bodo language as a medium of instruction, followed by the demand for the introduction of Bodo script in education in the 1970s. Their persistent movement compelled the Assam Government to grant Bodo language the status of an associate official language from 1984.[10] Thereafter, a movement for a separate State was spearheaded by the All Bodo Students' Union, a student organisation of Bodos. It mobilised tens of thousands of supporters against what they claimed 'Assamese chauvinism', which they considered to be the root cause of their alienation. It eventually formed the Bodo People's Action Committee as its political wing.

The movement came to an end after signing a peace accord in 1993 popularly known as First Bodo Accord between the government (Central and Assam Governments) and leaders of the movement to constitute an autonomous region within Assam to grant extensive home rule powers through a 40-member Bodoland Autonomous Council. The council consisted of a general council, comprising both elected and nominated members and an executive council who were elected from the members of the general council to carry out its executive functions. The council was entrusted with a certain amount of legislative, executive and financial powers on various issues – cottage industries, forests, cultural affairs, irrigation and so on. The council's annual budget was allocated by the Central Government in consultation with the Assam Government. An interim executive council was constituted in May 1993 which led to several Bodo Volunteer Force militants to lay down arms. The leaders of the movement floated the Bodoland People's Party, a political party. Unfortunately, failure to demarcate the exact boundary of the autonomous region coupled with an intense infighting among the leaders derailed all the efforts and hence led to the revival of another phase of the movement once again.

Second Bodo Accord, 2003

The resumption of the statehood movement was also resurrected armed conflict led by the Bodo Liberation Tigers, constituted by former Bodo Volunteer Force militants in 1996. It denounced the first accord and demanded a separate State. Subsequently, it also started negotiation with the government and agreed to give up the demand for a separate State and accepted limited autonomy with extensive home rule powers. Thus, another peace accord popularly known as Second Bodo Accord was signed in 2003 with limited autonomy within Assam under the Sixth Schedule of the constitution. The region is now technically known as the Bodoland Territorial Areas District (BTAD) comprising four contiguous districts of Kokrajhar, Baska, Udalguri and Chirang. The purpose is to fulfil economic, educational and linguistic aspirations and the preservation of land rights, sociocultural and ethnic identity of the Bodos and to speed up the infrastructure development.[11] The region is governed by a 40-member Bodoland Territorial Council, headquartered at Kokrajhar, an important town of the region. The council has a general council with 40 elected and six nominated members, and a 12-member executive council. The council has legislative, executive and financial powers in respect of over 40 subjects (Nath, 2003). Although the Bodo Liberation Tigers-led armed conflict had ended a decade ago, the ethnic conflict has not completely subsided due to ethno-religious tensions between the Bodos and the Muslims in the proposed area.

Argument and findings of the study

If we borrow the concept of negative peace introduced by Johan Galtung, peace accords constitute a significant move towards conflict resolution and restoration of 'negative peace' in Northeast India. If the government is unable to fully resolve the root cause of armed conflict in the region, it can nevertheless keep the activities of its opponents to a sufficiently low level through political reconciliation and use of military force. So far, the government has been successful in convincing its adversaries to accept its terms and conditions and more or less came forward for dialogue within the framework of the law of the country. Therefore, peace accords are the instruments through which the State imposes its will on the body of politics while trying to work out compromise on contentious political issues (Rupensinghe, 1996). Despite a number of peace accords, barring Mizoram (considered as one of the successful accords), not much respite of conflict is visible in the region. What is wrong in it?

In the case of Mizoram, the peace accord turned out to be successful at least in bringing 'negative peace' as noted by Johan Galtung. The MNF was the sole armed group and Laldenga was the most acceptable leader among the Mizos at that time. When the negotiation temporarily broke down in 1982, the religious leaders asked the militants to desist from violence which encouraged the resumption of the dialogue (Das, 2007). In the words of Zoramthanga, a former rebel leader, 'NGOs, Church leaders, political parties and groups of society asked the underground MNF to have peace talks with the government. We did accordingly and the peace accord was signed' (Malsawami, 2003: 144). Since the militant group was united under Laldenga, the peace accord was welcomed by a large section of the population. Due to his able leadership, 'the huge numbers that gathered to participate in Laldenga's funeral in Aizawl in 1990, it was a tribute to his charisma and the courage and statesmanship he deployed in the end in bringing back his followers to the constitutional fold' (Verghese, 1996). Though the Mizo Accord – 1986 was often described as one of the success stories in the country, but, in reality, it was settled by oppressive manner, following the Indian Air Force bombed Mizoram (Ngaihte, 2013). Besides, thousands of Brus and Chakmas have been expelled into Tripura in 1997, and militant groups within both minority communities have been crushed, in part by Mizo Youth Organizations acting as an extra-legal police force (Lacina, 2009).

In the case of the Bodos, after the formation of the Bodoland Autonomous Council in 1993, their leadership was marred by intense factional struggles. The situation was exploited further by the Assam Government by using one faction against the other. For instance Bwismutiary, the first Chairman of the council, could not discharge his duties due to indifferent attitude of the Assam Government. As a result, the administrative boundary of the council could not be demarcated. The Assam Government had agreed to include altogether 2,570 villages instead of the 3,085 villages demanded by Bwismutiary, who in protest, resigned from the chairmanship. The government appointed another factional leader, Prem Singh Brahma, since the latter agreed to accept the government's conditions (Hazarika, 1995) and his appointment further divided the Bodoland People's Party, the political party. Consequently, a faction of the Bodoland People's Front led by Bwismutiary started a movement for a separate State alleging failure on the part of the government to implement the provisions of the peace accord of 1993. Since the council's territorial boundary remained unresolved, the routine elections could not be held. This led to the feeling among the people that only a separate State would serve their purpose. From this, a greater question arises is that had the territorial boundary of the autonomous region been fixed before concluding the accord and had the Bodo Volunteers Force militants have been rehabilitated, the Bodo Liberation Tigers would not have been formed. Thus, one of the starkest features of this accord was that both the government and the Bodo leaders failed to convince hard core militants to surrender (Hazarika, 1995).

Contrastingly, the second accord of 2003 offered a larger autonomous region with extensive home rule power. This peace accord has not fully satisfied them but the conflict spearheaded by the Bodo Liberation Tigers has ended. The accord of 2003 opened up series of violence, rather than taking care of the non-Bodos' fear including the violent conflict between Bodos and Muslims. The Bodos who believed themselves as the earliest inhabitants in Assam have been apprehensive of immigration of numerically dominant ethnic groups. They alleged that a large number of illegal immigrants have settled in lands belonging to them leading to land alienation from them. The violent conflict between Bodos and Muslims in 2012 had claimed over eighty people and displaced over four lakh. The Bodoland People's Party, the ruling party in the BTAD, alleged that it was a conspiracy to destabilise the region. The party had alleged that it was incited by 'foreigners' and accused illegal migration as the root cause. On the other hand, the All India United Democratic Front, another political party, accused the Bodoland Territorial Council of promoting violence against the Muslims. It alleged that the violence was aimed at driving out non-Bodos from the BTAD. The party also alleged that the Bodos constitute about 29 per cent of the total population of BTAD. In the May 2014 Lok Sabha elections, the consolidation of non-Bodo votes ensured Naba Kumar Sarania's win by the biggest margin from Kokrajhar, the heart of Bodo politics. Sarania defeated his nearest independent candidate UG Brahma by a margin of over three lakh votes. Sarania was an independent candidate backed by Janagostiya Aikhya Manch, an organisation of the non-Bodos. Since the autonomous councils are named after the majority tribe residing there (e.g. Bodo), on the basis of their historical origin, assert their identity through coercive ways, and constantly resort to ethnic cleansing to demonstrate their control and authority over the region (Deori, 2013).

The largest issue remains – the peace accords with the Nagas. After the failure of the 9-Point Agreement in 1947, the 16-Point Agreement was concluded with the NPC without the involvement of any armed group. It was possible since the government believed that the armed groups would lose relevance once a State came into existence. But, the militants rejected the agreement (Nag, 2002). The formation of Nagaland was strategically planned to divide the Nagas (Ao, 2002; Nuh, 2006). The agreement divided the Naga political class and hence led to the emergence of an alternative political platform, from which the secessionist campaign could be politically challenged by the Nagas loyal to India (Bhaumik, 2007). In this context, Iralu (2003) said the agreement is the greatest betrayal in Naga history. Again, in 1975, the government succeeded in convincing six Naga leaders believed to be close to some militants of the NNC to sign another peace accord, popularly known as the Shillong Accord, with 'expected confusion' (Chaube, 1999) in which they agreed to abide by the Constitution of India and surrendered arms. In reality, in these processes, the government gave nothing to them. Subsequent to the signing of the accord, the NNC divided into two, in which the breakaway faction founded the National Socialist Council of Nagaland (NSCN). It got further bifurcated in 1980 into Khaplang-led group (NSCN-K) and Isak and Muivah-led group (NSCN-IM). Soon, the two factions used violence means to liquidate each other. Both groups have also engaged in a war of words, especially through the local media. Nonetheless, the Isak–Muivah-led group has emerged as the more powerful force, but others too have their share of influence.

In nutshell, the 16-Point Agreement and the Mizo Accord led to the formation of two new States of Nagaland and Mizoram, respectively, carved out from Assam. The two peace accords with the Bodos have led to the formation of an administrative region with Assam with limited autonomy, now known as BTAD. In the Shillong Accord, the government merely promised that its adversaries 'should have reasonable time to formulate other issues for discussion for final settlement'.[12]

The three ethnic groups, the Mizos, the Nagas and the Bodos, have not achieved their goals fully by the peace accords. The Mizos, who fought for independence from India, got a State. The Bodos, who waged armed struggle for a separate State, got only a limited political and financial autonomy. The Nagas, whose original demand was independence from India, had to compromise with the State of Nagaland (Ravi, 2012). It has thus underscored the fact that the government would not entertain the demand for secession. Once the militant groups and other leaders realised that the secession is impossible, they worked out for an alternative or second best settlement, which is the creation of a new State or an autonomous region. This was because the leaders of the ethnic groups viewed the State Government as an instrument by which to extend, consolidate or transform their position in the economy and social system (Weiner, 1983). It is because ethnic groups compete intensely to control State/political power and exert significant cultural influence. In this process, the numerically smaller ethnic groups have united for political purposes in their competition with a numerically dominant ethnic group so as to make them advantage politically and culturally. Such identity is termed as manufactured ethnic identity (Prabhakara, 2005).

As the government makes different commitments to different militant groups, it is unlikely to reach the situation of Zartman's Mutually Hurting Stalemate in the region. New armed group may be formed when the effort was made to control the older ones, probably due to the indecisive and the divisive policies of the ruling class (Ravi, 2012). For instance, the Naga militants were excluded when the Central Government negotiated with the NPC which led to the formation of Nagaland in 1963 (Fernandes, 1999).

Concluding remarks

Several decades of armed conflicts in Northeast India have compelled the conflict parties to realise the importance of negotiated settlements. Of the five peace accords discussed, only one was seemingly fruitful (as perceived by general public), that is Mizo Accord. In the case of other accords, instead of leading towards the resolution of conflict, most of them have worsened the situation of conflicts. Ignorance of the government officials and 'take it easy' approach of the state caused conflicts in the region, and non-inclusion of major stakeholders/parties is identified as an important factor for the failure of these peace accords. It is also evident from the facts that the strategies suitable in one conflict environment may not work in another environment. Strategies should be worked out on a case-to-case basis. The successful peace accord depends on the availability of popular leaders, but most armed groups and ethnic leaders are factional ridden. As a result, peace accord arrived at with one faction is often opposed by other factions. In this regard, the role of the government is crucial. If the government succeeds in working with all the factions together, the outcome will be positive. Thus, it is important to make dialogue with all factions rather than one or two. If not, the government must be willing to negotiate with major groups/factions which enjoy popular support, and can influence community. Otherwise, the outcome of signed treaties will be counterproductive. Rushing for peace accords without proper ground work must be avoided. In the hush-hush accords, the substantial issues are not properly discussed.

Past experiences have shown that the government was always opened to negotiate with any 'willing' groups without assessing their relevance. This gives the impression that government is interested only in making peace accords, one after another, without assessing their long-term impact. Monitoring of the implementation of peace accords also requires special attention. It however seems to be neglected in the past. Further, once a peace accord is signed, the particular armed group shall be encouraged to join in electoral politics. Finally, Northeast India has vocal civil societies working on human rights and other societal issues. Beyond that, it is the right time now to nurture them in a way that it will facilitate to fill the gap that has been created between government and armed groups.

Notes

1 India is a federal union and its constituent units are known as 'States'. The NER comprises Arunachal Pradesh, Assam, Manipur, Meghalaya, Mizoram, Nagaland, Sikkim and Tripura.
2 They are Bangladesh, Bhutan, China, Myanmar and Nepal.
3 This is the definition adopted by the Uppsala Conflict Data Program, Uppsala University, Uppsala.
4 This is the definition adopted by the Uppsala Conflict Data Program, Uppsala University, Uppsala.
5 It provides for autonomous district and autonomous regions within those districts with elected councils which enjoy powers to levy taxes, to constitute courts for the administration of justice involving tribes and law-making powers on land allotment, occupation or use of land, regulation of '*jhum*' or other forms of shifting cultivation, establishment and administration of village and town committees and the like. See, for example, Singh, M. Amarjeet (2007). 'Challenges before Tribal Autonomy in Assam', *Eastern Quarterly* 4(1): 27–35.
6 Detail of this section can be seen from Singh (2012).
7 They were I. Temjenba, Dahru, Veenyiyl Rhakhu, Z Ramyo, M Assa and Kevi Yallay.
8 It was appointed by the Central Government in 1953 to examine any proposal for states reorganisation in the country.
9 Union Territories are administrated by the President of India acting to such extent, as she/he thinks fit, through an administrator appointed by her/him.
10 Each state in India can enact its official language. The Bodo language has been an associate official language of Assam for specific purposes. It has also been recognised as the official language of the BTAD.

11 See, The Memorandum of Settlement between the Government of India and the Bodo Liberation Tigers of 2003.
12 See, the full text of the Shillong Accord.

References

Adler, Emanuel (1998). 'Condition(s) of Peace', *Review of International Studies*, 24(5): 165–92.

Alberth, Johan and Carlsson, Hennin (2009). 'Critical Security Studies, Human Security and Peace' (D-level paper, Political Science), Linköping University.

Ao, Lanunungsang (2002). *From Phizo to Muivah: The Naga National Question in North-East India*, New Delhi: Mittal Publications, 81.

Arnson, Cynthia, J. (1999) (ed). *Comparative Peace Processes in Latin America*, New York: Stanford University Press, 1.

Bapat, Navin A. (2005). 'Insurgency and the Opening of Peace Processes', *Journal of Peace Research*, 42(6): 699–717.

Bartos, Otomar J. and Paul Wehr (2002). *Using Conflict Theory*, New York: Cambridge University Press, 13.

Bell, Christine and Catherine O'Rourke (2007). 'The People's Peace? Peace Accords, Civil Society, and Participatory Democracy', *International Political Science Review*, 28(3): 293–324.

Bhaumik, Subir (2007). *Insurgencies in India's Northeast: Conflict, Co-option and Change*, Washington: East-West Center, 11.

Boulding, Kenneth, E. (1978). *Stable Peace*, Austin, TX: University of Texas Press.

Chaube, S.K. (1999). *Hill Politics in Northeast India*, New Delhi: Orient Longman, 252.

Cline, Lawrene E. (2006). 'The Insurgency Environment in Northeast India', *Small Wars and Insurgencies*, 17(2): 126–47.

Das, Samir Kumar (2007). 'Conflict and Peace in India's Northeast: The Role of Civil Society', Working Paper No. 47, Washington: East-West Center, 1–3.

Deori, N. (2013). 'Ethnic Fratricide and the Autonomous Councils of Assam, Yojana', August 1 (Online Version). Accessed: http://yojana.gov.in/fatricide.asp.

Derek, Jinks (2003). *The Temporal Scope of Application of International Humanitarian Law in Contemporary Conflicts*, Cambridge: Harvard University, 3.

Fernandes, Walter (1999). 'The conflict in the North-east – A Historical Perspective', *Economic and Political Weekly*, 34(51): 3579–82.

Galtung, Johan (1967). *Theories of Peace: A Synthetic Approach to Peace Thinking*, Oslo: International Peace Research Institute, 6.

Harbom Lotta, Stina Högbladh and Peter Wallensteen (2006). 'Armed Conflict and Peace Accords', *Journal of Peace Research* 43(5): 617–31.

Hazarika, Sanjoy (1995). *Strangers of the Mist: Tales of War and Peace from India's Northeast*, New Delhi: Penguin Books, 162.

Heimerdinger, Philipp and Tshering Chonzom (2012). 'Conflict in North-east India: Issues, Causes and Concern'. Briefing Paper. New Delhi: Heinrich Böll Stiftung – India. Accessed 20 December 2013: http://www.in.boell.org/web/52–259.html.

Höglund, Kristine (2008). *Peace Negotiations in the Shadow of Violence*, Boston: Martinus Nijhoff Publishers, 14.

Inoue, K. (2005). 'Integration of the North-East: The State Formation Process', In Murayama, M., Inoue, K. and Hazarika, S. (eds), *Sub-Regional Relations in the Eastern South Asia – With Special Focus on India's North Eastern Region*. Chiba (Japan): The Institute of Development Economics, 16–31.

Iralu, Kaka D. (2003). '16 Point Agreement- The Greatest Betrayal in Naga History', *The Sangai Express*, February 25. Accessed 31 May 2013: http://www.npmhr.org/index.php?option=com_content&view=article&id=79:16-point-agreement-the-greatest-betrayal-in-naga-history-&catid=26:naga-peace-process&Itemid=91.

Jacob, Malsawami (2003). 'Mizoram's Peace March', In Vattathara, T. and George, E. (eds), *Peace Initiative: A North East India Perspective*, Guwahati: Don Bosco Institute, 144.

Johnson, L. Gunnar (1976). 'Conflicting Concepts of Peace in Contemporary Peace Studies', *SAGE Professional Paper in International Studies*, l(4), 2–46.

Lacina, B. (2009). 'The Problem of Political Stability in Northeast India: Local Ethnic Autocracy and the Rule of Law', *Asian Survey*, 49(6): 998–1020.

Nag, Sajal (2001). 'North East: A Comparative Analysis of Naga, Mizo and Meitei Insurgencies', *Faultlines*, 14. Accessed 31 May 2013: http://www.satp.org/satporgtp/publication/faultlines/Volume14/Article4.htm.

Nag, Sajal (2002). *Contesting Marginality: Ethnicity, Insurgency and Sub-nationalism in Northeast India*, Delhi: *Manohar Book*, 258.

Nath, M.K. (2003). 'Bodo Insurgency in Assam: New Accord and New Problems', *Strategic Analysis*, 27(4): 533–45.

Ngaihte, T. (2013). 'Manipur and Its Demand for Internal Autonomy', *Economic and Political Weekly*, 48(16): 20–1.

Nuh, V.K. (2006). 'The Cry of the Naga People', In Prasenjit Biswas and C J Thomas (eds), *Peace in India's Northeast: Meaning, Metaphor and Method*, New Delhi: Regency Publications, 245.

Prabhakara, M.S. (2005). 'Manufacturing Identities?', *Frontline*, 22(20): 20. Accessed 20 May 2014: http://www.flonnet.com/fl2220/stories/20051007002609500.htm.

Pruitt, Dean G. and Sung Hee Kim (2004). *Social Conflict: Escalation, Stalemate, and Settlement* (3rd Ed.), New York: Tata McGraw-Hills, 172–89.

Ravi, R.N. (2012). *Chasing a Chimeric Peace*, New Delhi: *The Hindu*, 15 November 2012. Accessed 15 November 2012: http://www.thehindu.com/opinion/op-ed/chasing-a-chimeric-peace/article4095592.ece.

Ray, Animesh (1993). *Mizoram*, Delhi: National Book Trust.

Rupensinghe, Kumar (1996). 'Strategies for Conflict Resolution: The Case of South Asia', In Kumar Rupensinghe and Kumar David (eds), *Internal Conflicts in South Asia*, London: Sage Publication, 180.

SATP (2012). 'Incidents and Statements Involving Black Widow 2004–12', New Delhi: Institute for Conflict Management. Accessed 30 May 2013: http://satp.org/satporgtp/countries/india/states/assam/terrorist_outfits/blackbw_tl.htm.

Singh, M. Amarjeet (2012). *The Naga Conflict*, Bangalore: National Institute of Advanced Studies.

Singha, Komol (2012). 'Identity, Contestation and Development in North East India: A Study of Manipur, Mizoram and Nagaland', *Journal of Community Positive Practices*, XII(3): 403–24.

St. John, Anthony Wanis (2008). 'Peace Processes, Secret Negotiations and Civil Society: Dynamics of Inclusion and Exclusion', *International Negotiation*, 13(1): 1–9.

Stedman, Stephen J. (1997). 'Spoiler Problems in Peace Processes', *International Security*, 22(2): 5–53.

Tiernay, Michael (2012). 'Battle Outcomes and Peace Agreements'. Accessed 2 June 2013: https://files.nyu.edu/mrt265/public/tiernay_battle_data.pdf.

Verghese, B. G. (1996). *India's Northeast Resurgent: Ethnicity, Insurgency, Governance and Development*, New Delhi: Konark Publishers, 151.

Weiner, Myron (1983). 'The Political Demography of Assam's Anti-Immigrant Movement', *Population and Development Review*, 9(2): 279–92.

Zartman, I. William (2001). 'The Timing of Peace Initiatives: Hurting Stalemates and Ripe Moments', *The Global Review of Ethno-politics*, 1(1): 8–18.

——— (2008). *Negotiation and Conflict Management: Essays on Theory and Practice*, New York: Routledge Publishers, 54.

Chapter 6

Forest conservation and community land rights in Manipur

Hoineilhing Sitlhou

Introduction

The Supreme Court of India's decision to let the *gram sabhas* decide the fate of Vedanta's Niyamgiri mining project in Niyamgiri hills of Odisha is a path-breaking move in which the consent of tribes and local population is taken by the government. The apex court's ruling put *gram sabhas* or village assemblies at par with statutory and regulatory bodies. The incident brings to memory the debates around the Scheduled Tribes Bill 2005. Under this Bill, 2.5 hectares of forest land is proposed to be given to each tribal family occupying forest land before 25 October 1980 (Munshi, 2005: 4406). The land is given only for livelihood and not for commercial purposes; it is inheritable but not transferable or alienable (Ibid.: 4407). Interestingly, 'the right to allot this land, which is to be registered jointly in the name of a male member of the family and his spouse, is given to the *gram sabha* of the village concerned' (Ibid.: 4406). The government is trying to establish a more democratic management of forest for forest communities through the agency of the *gram sabha*. This chapter attempts to apply this to the understanding of the relationship between the state, the forest and the Scheduled Tribes (STs) in Manipur with special reference to the Kuki community. The chapter argues that the same democratic management of forest which involves the local communities, their institutions and knowledge be encouraged in dealing with the hill communities of Manipur.

The word 'Kuki' is a generic term which includes a number of tribes and clans. 'Kuki' refers to an ethnic entity spread out in a contiguous region in Northeast India, Northwest Myanmar and the Chittagong Hill Tracts in Bangladesh (Sitlhou, 2014: xi). This research work focuses on the 'Thadou' dialect speaking group among the Kuki communities of Manipur. As per 2001 census, they constitute the largest ST population with 1.8 lakh and represent 24.6 per cent of Manipur's total ST population.[1]

Community land rights and reforms

Before the advent of British rule in India, the regulation of people's use of forest was mainly done through the local customs (Kulkarni, 1987: 2143). In theory, the entire land of a Kuki village belongs to the Chief. The chieftainship system is woven around the concept of privileges and obligations of both the Chief towards the subjects and vice versa. According to Ray (1990: 38), the Kuki polity is ordered into seven-tiered structures with the Chief or *Hausapu* as the head. He is assisted by the following officers:

- *Semangpa* (Prime Minister)
- *Pachong* (Secretary)
- *Thiempu* (Physician-cum-priest)
- *Thihiu* (Village blacksmith)
- *Chonloi* (Treasurer)
- *Lom-upa* (Youth Director in-charge of youth co-operative works)
- *Kho-sam* (Announcer of the decisions of the village council)

They have their own customary court which is the traditional law-enforcing body, comprising the Chief and his council of ministers. It is the highest body of law in the village and is governed by traditional customary laws that are unwritten and retained orally. A council of elders known as *Semang-Pachong* or village authority leaders are elected to assist the Chief in superintending and transacting all business matters in connection with the land cultivation, measurement, collection of tax and so on (Sitlhou, 2011a). This body is equivalent to the *gram sabha* in other parts of the country. The Kuki customary court also makes rules regarding forest laws. It has survived the onslaught of external agencies that have directly impinged on the society like colonialism and the democratic set-up of the Indian nation state. The customary court is a law-enforcing agency that preserves and uplifts the traditional customary laws within contemporary legal systems (Ibid.: 365).

In the hilly areas of Manipur, the community, not the state, owns most of the land. Land and forests are the two basic resources for subsistence in the hills. This implies that the villagers hold the land with the sanction of the Chief and community and not the government. Land-ownership exists under chieftainship system, but in a form quite different from other land systems. There are different types of proprietors of land in the village. There are those who procure permanent ownership rights to their land through the Chief because they had assisted the Chief at the time the village was established or had been gifted land for several reasons. In theory, the land is temporarily owned but these types of ownership assume a sort of permanency over years of residence. There are the temporarily allotted sites for jhum lands or kitchen gardens usually in the mountainside or vacant plots of the villages. This type of ownership is temporary and has to be renewed after a period fixed by the Chief and the council of ministers (Sitlhou, 2011b: 329). The Chief is therefore endowed with 'the rights of management of community resources and in the exercise of this he is authorised by tradition' (Burman, 1992: 275).

Manipur has two distinct physiographic divisions of the hills and the plains. It is nine-tenths hills and one-tenth plains (Das, 1995: 48). Manipur has a total area of 22,327 sq km out of which the valley covers an area of 2,238 sq km and the total population is 2,166,788.[2] The hill area covers nine-tenths of the total area of the state, which indicates that the tribal areas are scarcely populated with a density of 44 persons per sq km. The density in the valley is 631 persons per sq km (Kipgen, 2009: 329). The ownership of land in Manipur is a combination of community and private ownership of land. Ownership in the valley is acknowledged by patta system certified by the Deputy Commissioner. Whereas, in the hill areas under chieftainship, the legitimacy of ownership rights over land is given by the Chief. The subdivision officers or district magistrates, under whose jurisdiction the village land belongs, issue documents, which are considered to be equivalent to patta in the valley (Gangte *et al.*, 2010: 132). This was given cognizance by the British government.

The post-independence era however witnessed a succession of legislations which were meant to administer the hill areas in the name of development and uniform policy of the state. All the legislations have many erosive effects upon the authority structure of the traditional chieftainship system and on the land rights of the community, both directly and indirectly. The Manipur Hill Peoples Administration Regulation Act, 1947, wanted to do away with the traditional tributary privileges of the Chief, whereas the Manipur (Village Authority in Hills Areas) Act, 1956, attempted to reduce the status of the Chief to that of the ex-officio chairman of the village authority. The Manipur Hill Areas Acquisition of Chief's Rights Act, 1967, directly attempted to abolish chieftainship in the hill areas by paying compensation (Haokip, 2009: 313).

However, the most controversial act was the Manipur Land Revenue and Land Reform Act of 1960. As per the Act, all land including forests, mines and minerals are not the property of any person but of the state. It was first directed only at the valley areas and had special provisions for safeguarding the land rights of the hill communities. 'The MLR and LRA, 1960 was sought to be amended through the Sixth Amendment Bill (1989) by repealing the provision which (a) excluded the hill areas of Manipur from the ambit of the Act and (b) protected tribal land under section 158'.[3] The proposed amendment bill inopportunely suggested the deletion of the special safeguards against land alienation.

In most parts of the country, the Land Reform Act was introduced with any one (or all) of the following objectives: (a) implementation of ceiling regulation, (b) introduction of tenancy reforms and (c) removal of intermediaries (Ibid.: 229). The primary objectives include preventing low productivity communitarianism, promoting investment and growth, developing a land market and encouraging 'individualisation' (Ibid.: 228). The term 'intermediary' is slightly problematic when contextualised in the case of Northeast India because of its earlier preoccupation with exploitative agencies like zamindars in mainland India. In the case of Northeast Indian societies, the source of livelihood is not only cultivation but also forest products. These communities are extremely dependent on the forest for fuel, fodder, shelter, fruits, timber, employment and other products and have access to these resources via their membership in the community which is held together by the institution of chieftainship system. 'Hence, if the community as the "intermediary" between the state and individual householders is removed, the tribal households will be actually dispossessed of much larger resources, though their rights over small areas of settled agricultural land will be consolidated. They, therefore, fear that along with individualization, they may not have access to community forests which will pass into the hands of the state . . .' (Ibid.: 229). Moreover, this will allow for land revenue exactions by the state and facilitate the large-scale plunder of timber and other forest wealth. Therefore, protests have been held calling for preserving the customary tribal land ownership institutions in diverse methods and movements in different periods of time, restoration of chieftainship system demand for Sixth Schedule and collective agitations against uniform policy.

Traditional ecological knowledge and sustainable development

Traditional ecological knowledge (TEK) represents a collective understanding attained over long periods of time, in particular places, of the relationship between a community and the earth. It may encompass spiritual, cultural and social aspects as well as substantive and procedural ecological knowledge. It may also include customary rules and laws, rooted in the values and norms of the community to which it belongs. The Kukis' reverence for nature is reflected in the sacred space and institution it occupies in the society as also the accompanying rituals credited to it. The centrality of ecology in the cultural life of their society is evident from ecological sensibilities that are ingrained in their cultural practices and their understanding of the universe. For instance, the naming of the twelve months of the year for this community is in accordance with the seasonal changes in nature and the position of the moon in the sky (Sitlhou, 2011).

Land rituals: respect and reciprocity

The discourse between nature and man is made evident in agricultural rituals of the Kukis. They performed several ceremonies such as purification of forest when choosing a site (*lou-mun-vet* and *daiphu* rituals), purification of soil after slashing and burning down the forest, ritual to invoke blessing and good fortune (*Chang-Nunghah* ritual), thanksgiving (*Chang-ai/Sa-ai*) or the harvest festival (*Kut*). Therefore, each stage in the agricultural calendar has its own accompanying rituals. In case a natural water-spring happened to exist in a particular plot of land, the owner is required to perform the *twikhuh thoina*. A request is made to the spirit of the water to reside seven steps under the earth by gifting them symbolic elements.

In all the rituals, the priest acted as a middleman between the people and the forest spirits to facilitate their discourse and negotiations. The rituals also reflected the worldview of a sense of reciprocity and respect towards the other creations in the cosmos. The local people knew the need for a built-in compensation for human actions or some act of 'reciprocity' (Tinker, 2008: 68). The ritual for purifying land and selection of site for cultivation (*lou-mun-vet ritual*) showed the sensibility and sensitiveness of the people towards the other beings in the cosmos. Permission was sought from the spirits who were believed to be the previous occupants and there was the observance of a day's curfew as a symbolic expression of condolences for all the animals and insects that perished in the jungle fire while clearing the field. 'Thus, nothing is taken from the earth without prayer and offering' (Ibid.: 70). All the rituals involved a process of negotiation with the spirits who are thought to be earlier possessors of the land. They were also attempts to appease spirits that were believed to have a temporal authority over the land and its produce (Sitlhou, 2011). The belief system and mystical values transformed the soil from a merely physical entity into a culturally determined object (Ibid.).

Thus, the respect for the soil, the land or nature came naturally in the cosmology of the Kukis and this defined the human–environment interactions. The land was understood to be both productive in its utility and also a space of spiritual communion and moral structure.

Customary governance systems

The rights in the soil and the land tenure systems are different under the caste system and the chieftainship system of the tribals. The land system and settlement pattern in most Indian villages are governed by the rules of hierarchy under the caste system which in turn influences the distribution of rights and privileges (as cited in Mandavdharc, 1993: 78). In kuki villages, the allocation of jhum land, site for settlement within the village, access to forest area and the regulation of the use of rocks, water, soil, woods, grass and other non-timber forest products are done by the Chief assisted by the council of ministers. This is where village institution of governance plays an important role in regulating and mediating between human and nature (Sitlhou, 2011).

Traditionally, as per their custom, the *Haosa* or Chief has absolute power over the village land and it is his duty to distribute the cultivable land to all the villagers at the beginning of every year. In doing so, he has to consult his council of ministers called *Semang-Pachong* nominated by him from different clans of the village. It is obligatory for any hunter to offer to the *Haosa* the head and the right hind-leg of any animal killed by the hunter. The villagers are obligated to give the Chief a basketful of grains of rice called *chang-seu* at the end of the year (Devi, 2006: 52). This is done as an act of acknowledgment that the Chief is the overseer of all the lands and all the produce that comes of it (Sitlhou, 2011).

New settlers in order to get homestead land have to get the permission of the Chief after traditionally serving *ju* or rice-beer to the Chief. With the advent of Christianity, this has been replaced by '*cha-omna*' or 'serving of tea'. Gifts like cocks, shawls or vegetables could be given alongside at this time of negotiation. The land allotted to the villagers cannot be sold. If a family in the village wishes to migrate to another village, the land will automatically be returned to the Chief. The members of the Chief's council with the approval of the Chief superintend transact all business matters in connection with the land – cultivation, measurement, collection of tax and so on. When a particular land is to be cultivated for jhumming purposes by a villager, it has to be brought to the knowledge of the Chiefs for permission.[4]

Therefore, the customary governance system tries to maintain the balance of nature and protect the territory as a whole against the large-scale exploitation of the forest. Judicious use of the forest resources is facilitated under the chieftainship system.

The forest act in India and British legacy

A systematic forest policy in the country was begun in 1855 when the then Governor General Lord Dalhousie issued a memorandum on forest conservation (Kulkarni, 1987: 2143). Under the Forest Act of 1878, the forests were divided into (1) reserved forests, (2) protected forests and (3) village forests (Ibid.). These regulations were formally initiated in 1894 (Anderson and Huber, 1988: 36). Anderson and Huber's explanation of the implication of the forests division is relevant to our study, in their words:

> Reserved forests were exclusively for the use of the Forest Department except for certain minor concessions, such as gathering of the fruit of the trees and cutting of the grass, on payment of small dues. In the reserved forests, the surrounding villagers had no rights other than the ones explicitly permitted by the state. The protected forests were also managed by the Forest Department, but the people of the surrounding villages had certain rights in them, such as gathering fruits and other produce of the trees, and cutting timber and wood specifically for the use of the villagers (but not for sale). They also had freedom to graze their livestock and hunt wild game for domestic purposes. Over the protected forests, the villagers had all rights not specifically taken away by the state. The village forests were the communal property of the villagers. (Ibid.: 37)

The Indian Forest Act of 1878 allowed the state to increase the commercial exploitation of the forest while putting curbs on the local people's access to the forest for subsistence. This denial of forest rights to the villagers provoked countrywide protests. In 1988, the forest policy was changed to be more accommodative towards the local people (Saxena and Sarin, 1999: 187). The new policy replaced the earlier focus on commercial exploitation of forests with the twin objectives of maintaining ecological stability and meeting the forest-based needs of forest dwellers and other rural poor living in and near forest areas (Ibid.: 187). The emphasis of the new policy was to protect the legal rights, concessions and privileges of tribal and other local villagers. The new policy encouraged the involvement of local women and men in the protection, management and development of forests (Ibid.: 187).

The Indian Forest Act of 1927 was an attempt to regulate the rights of the people over forestland and produce as the government gradually increased its control over the forests by strengthening the forest department (Kulkarni, 1987: 2143). Subsequently, there was a steady and considerable increase in revenue obtained from the forests (Ibid.: 2143). After Independence, there was some rethinking on the issue of forest policy in putting the national needs over the claims of the communities living in and around the forests (Ibid.: 2144). The National Forest Policy was issued in 1952 with this in mind. 'Though the traditional rights of the tribal were no longer recognised as rights, in 1894, they were declared in the British Forest Policy as "rights and privileges". In 1952, the National Forest Policy was further diluted into "rights and concessions". Now, the tendency is to treat them merely as "concessions"' (Burman, 1992: 143). The subject of forests was included in the state list of Seventh Schedule in the Constitution of India (Kulkarni, 1987: 2144). There was a major change which took place in 1976 and the subject of forests was transferred from the state list to the Concurrent list through the 42nd amendment to the Constitution (Ibid.: 2144). Power over forests was transferred from the control of the state to the centre; and the Government of India in the Forest Conservation Ordinance promulgated this in 1980 which later became a bill (Ibid.: 2144). The Indian Forest Bill, 1980, besides including a provision for curtailing the rights and benefits of the local people over the forests prescribes severe punishments for forest offences (Ibid.: 2144). Therefore, the revenue generation of the colonial period (1865 onwards) continued in the post-colonial period.

Additional responsibilities were added when social forestry was introduced in 1976, giving forest departments roles in promoting forestry on private and revenue land. Joint Forest Management has (since 1988) been a further element in the expanding remit of the forest bureaucracies. The signposts in this history are provided by the forest acts of 1878 and 1927, the Forest Policy of 1952, the National Commission on Agriculture of 1976, which inaugurated social forestry, the Forest Conservation Act of 1980, the 1988 Forest Policy and the GOI Joint Forest Management Resolution of 1990, followed by different state resolutions (Sundar and Jeffery, 1999: 27).

Many of the acts today take a strong conservationist stand against environmental degradation by restricting peoples' rights to the forest. The Scheduled Tribes Bill (Recognition of Forest Right) of 2005 recognises the right of the villages which are recognised under any state laws of any Autonomous District Council which are accepted as rights of tribals under any traditional or customary law of any state. The Act was criticised as it was considered to be detrimental to the goal of the National Forest Policy of 1988 that looks at bringing one-third of the country under forest (Munshi, 2005: 4406). The Conservation of Forests and Natural Ecosystems Act, 1994, seeks to meet its objectives by severely restricting local communities' rights to the forest. The Act proposes strict state control for 'sound ecological management' (Baviskar, 1994: 2494). Moreover, sections 3(2) and 27A of the Conservation of Forests and Natural Ecosystems Act vested the centre with the power to direct any state government to constitute reserved forests in any specific area. According to Ramachandra Guha, the Act will sharply curtail people's rights by classifying most forests, especially the better quality ones, as reserved forests, to be owned and managed by the state (Ibid.: 2494).

Forest policies in the Kuki villages in Manipur

The present research will analyse the case of two villages that are dominated by the Kuki group of people. The two villages – Motbung[5] and Tujang Vaichong[6] that come under the Sadar Hills subdivision of Senapati district of the state represent an interesting case of the evidence of both elements of tradition and modernity. The two villages are typical traditional Kuki villages which are under chieftainship. The local inhabitants of Tujang Vaichong and Motbung have also experienced the implications of the various forest policies of the government. Changes are evident in the land laws of villages especially in the case of forest laws in which the Chief and his council of ministers or village council incorporated the national laws of the country in their formulations of local law. Mrs. Konkhochong Kipgen, acting Chief of Tujang Vaichong village, says that it is the hereditary Chief who has sentiment towards the land and the people of the village. Commercialisation in terms of forest produce is checked to a certain extent in the jurisdiction of the chieftainship system.

The land laws in Tujang Vaichong include certain restrictions regarding forest laws. There are three types of forest areas – open reserved area, protected area and village area. This classification of forest has been made according to the Forest Act of 1878, which is still practiced in the village. The Chief opens some areas every year to the villagers for *jhumming* or *thinglhang lei* and for cutting woods in the mountains. The villagers have to seek the Chief's permission to lease out the land to them. They have to bring with them a rooster as a token or this is also done through the traditional *cha-omna* (where tea is served to the

Chief before a request is put forward). The land is reverted back to the Chief in case the occupant migrates to other village(s). The permission of the Chief is essential for procuring land for cultivation, taking woods from the forests and for cultivating in the mountain region. In Tujang Vaichong village, the permission granted by the Chief for land accessibility is valid for a year and has to be renewed every year. For cutting woods, the Chief opens the forest to the villager for about two to three months (mainly in the month of December, January and February). During that period, the villagers have to collect wood enough to sustain them throughout the year. In exceptional cases, a new settler can get access to the woods to build his or her house at any time of the year. However, there is a restriction against the use of these woods for commercial purposes. In relation to land ownership, there are five types of people in the villages. They are:

1 Landlord/Chief[7];
2 Landless labourers;
3 Those who cultivate their own land;
4 Those who lease out their land; and
5 Absentee landowners.[8]

Moreover, the settlement laws regarding those who are residing in the homestead area are obliged to:

- dutifully pay the *changseo* or village fund traditionally in the form of a basket of paddy or in cash[9];
- abide by laws laid down in the villages meant for the villagers; and
- neither to steal nor distort law and order in the village.

The house tax for every house is collected annually by the Chief and submitted in bulk to the government through the subdivisional officer (SDO) concerned. In 2008–09, an amount of rupees fifteen (Rs 15/-) was paid as house tax by each household.[10] The homestead area falls under the category of 'village area'.

The main land laws[11] in Motbung village are – first, annual tax of rupees 50 per house or 1 basket of paddy (*changseu*). Second, the forest is divided into two areas: (1) Protected Forest Area where no one is allowed to cut wood. For violating the laws in the protected area, there was an instance where a villager was fined *voh-cha*[12] or 'pig' for encroaching and violating the rules of the protected area. He had cut down some trees for his personal purpose. The verdict of the customary court is supreme and the penalty is finalised according to the norms prescribed by the customary law. The Protected Forest Area comes under the protection of the Chief of the village. (2) The second type is the Open Reserve Area where villagers can carry out certain activities like jhumming, grazing and firewood collection. Also, the villagers are allowed access to firewood, grazing, woods for new village settlers and jhumming (under the direction of the Chief who specifies the site for the year cultivation). The non-tribals (mostly Nepali migrants) pay grazing tax whereas for other villagers, this is covered under the annual tax or *changseu*. There is also a large reserved area controlled by the forest department under the office name, 'Range Forest Officer, Motbung Range'. These areas are strictly made inaccessible to local inhabitants.

In the case of the two villages – Tujang Vaichong and Motbung, it is only in 'village forest areas' that villagers, under the directive of their Chief, have adequate access to the forest. The government laws existed side-by-side with the local customary laws of the villages and towns. So, there is a two-fold authority system curtailing the local people's movement in the forest, which makes the forest and its resources' accessibility almost impossible. Forest departments are not operational effectively in most villages but they still have paramount legalised rights in lieu of the power bestowed upon them by various legislations. The classification of the village forest into three categories and the exclusive rights of the forest department over the 'reserved forests' under the Forest Act of 1878 is still effective in both villages. The Indian Forest Act of 1927 further increases the control of the government over forest in order to obtain revenue. This strengthening of the forest department was accompanied by alienation of the local people from the forestland and as per the National Forest Policy of 1952, the term, 'rights and privileges' of forestlands was gradually changed into 'concessions'. Moreover, sections 3(2) and 27(A) of the Conservation of Forests and Natural Ecosystems Act vest the centre with power to direct any state government to classify any area of the forest under the category of 'reserved forests'.

Regulations on practice of shifting cultivation

As a part of the forest conservation drive, the colonial land revenue settlements and forests laws were intended to curb and finally eradicate shifting cultivation (Sharma, 1994: 143). The Forest Act of 1865 was made to regulate forest exploitation, management and preservation (Ibid.: 143). 'For the first time an attempt was made to regulate the collection of forest produce by the forest dwellers. Thus, the socially regulated practices of the local people were to be restrained by law' (Ibid.: 143). The forest regulations sought to completely prohibit the practice of shifting cultivation in the central provinces (Ibid.: 143).

Recent debates have posited shifting cultivation or jhumming under two main perspectives – as the natural way of life of the tribal people and as detrimental to the forest economy. Shifting cultivation is defined by many as the natural way of life of some tribal people and the natural source of earning their livelihood (Bhowmick, 1980: 134). For others, it is detrimental to forest economy and therefore, to national economy as it leads to destruction of forest due to erosion of soil (Ibid.). The supporters of jhumming are of the view that it is more than sustenance and rather reflects the reason for existence. Shifting cultivation is deeply rooted in the Kuki psyche, having evolved through generations, and being rooted in customs, belief and folklore. It influences the tribe's mindset and cultural ethos of its agrarian society (Singsit, 2010: 158). Verrier Elwin was of the opinion that the tribes must be assisted to come to terms with their own past so that their present and future will not be a denial of their past but a natural evolution from it (Elwin, 1964: 302).[13] The ban on shifting cultivation is destructive to the mode of earning a livelihood as it failed to provide an alternative mode of livelihood (Savyasaachi, 1991).

In Manipur, soil conservation and land use programmes initiated to control or reverse the deleterious consequences of jhum are carried out by two agencies, viz., the department of horticulture and the forest department.[14] The voice from the educated section of the locals also resonates the ecological consequences of jhumming and its irrelevances today since existing population density far exceeds the carrying capacity.[15] In the framework of linear historical development and normative order of industrial production, 'social formation progresses from simple to complex, primitive to modern, savage to civilized, and irrational to rational. Thus, jhum cultivators are historically backward, and those who work with thermal projects are historically advanced' (Savyasaachi, 2001: 80). The tradition of shifting cultivation is thought to be in conflict with the tradition of modernisation and development (Ibid.: 83).

Conclusion

The existence of the dual system of administration regarding land rights in Manipur has led to a lot of complexities especially in the hill areas. The state agencies imposed laws without a real understanding of the ruled. The people have construed these laws to be indirect attempts to slowly erode the land ownership of the hill tribes in the name of a uniform land policy. The hill areas and the valley areas have their specific and distinct systems of land ownership and tenure. The diversities and differences of regions and cultures need to be understood primarily. This would require a series of consultations and studies to enable the formulations of separate but appropriate acts for different areas.

The villagers of Tujang Vaichong and Motbung have very limited access to the forest due to the existence of the two-fold authority system – customary laws and government laws – regulating activities on forestland. The Central Government's forest policies have increasingly alienated the villagers from forests which they could earlier freely access albeit within the boundaries of their customary laws. The forest laws which are meant to protect the forest and its resources alienate the local inhabitants who have been caretakers of the forest, and know the forest better than the state officials. In many cases, these forest laws are made without consideration of the 'ways of life' and 'worldview' of the governed. Instead of curtailing the people's rights to the forest whose livelihood depends on it, it would be more beneficial to design a forest management strategy that involves the local people. The TEK and the customary governance reflect the attitudes of the Kuki community on nature, a sense of both respect and desire to maintain a balance in the ecological structure. This worldview provides an in-built mechanism for environment protection. The state can work on enhancing, tabulating and using these knowledge and aspirations to involve the locals in the process of environment preservation instead of imposing land laws that are irrelevant. The local customary court and the District Council need to be acknowledged and empowered and can be used as a mechanism to have a dialogue and instil participation of the local people. Forest conservation is possible only if people's rights are recognised and established within a larger goal of their overall development and participation.

Acknowledgement

I would like to thank two special people who inspired and helped shape this article namely, Prof. Susan Visvanathan and Dr. Laldinmoi Pangamte.

Notes

1 Manipur (Data Highlights: The Scheduled Tribes) Census of India 2001, Accessed 13 February 2014: http://censusindia.gov.in/Tables_Published/SCST/dh_st_manipur.pdf.

2 Census of India 2001, Office of the Registrar General & Census Commissioner, Ministry of Home Affairs, GOI, Accessed 14 May 2011: http://www.censusindia.gov.in/Tables_Published/A-Series/A-Series_links/t_00_003.aspx.
3 Section 158: Special provision regarding Scheduled Tribes: No transfer of land by a person who is a member of the Scheduled Tribes shall be valid unless (a) the transfer is to another member of the Scheduled tribes or (b) where the transfer is to a person who is not a member of any such tribe, it is made with the previous permission in writing of the Deputy Commissioner, provided that the Deputy Commissioner shall not give such permission unless he or she has first secured the consent thereto of the District Council within whose jurisdiction the land lies or (c) the transfer is by way of mortgage to a co-operative society.
4 Kipgen, Konkhochong, Acting Chief (on behalf of her son) of Tujang Vaichong village, interviewed on 9 November 2008.
5 Motbung village is located on the sides of National Highway 39 in Manipur, which is about 26 km on the north from Imphal, the capital city. Chieftainship system is still functional and the village council consists of fourteen members, who are loosely elected by the community from the village itself. Women are not present on this traditional council. As per 2001 census, the village has about 478 households, but as per the chief census of 2008, the household population was 609.
6 Tujang Vaichong under Imphal–Tamenglong Road is under Kangpokpi Police Station. Tujang Vaichong is around 40 km away from Kangpokpi, and Kangpokpi is around 50 km north of Imphal. It is in the borderline between Senapati and Tamenglong district. It is a village with a household population of 156 as per the government census of 2001. The number of households as per the Chief's record in 2008 was 194.
7 Overseer of the whole land of the village.
8 A question was asked to the respondent on the relationship to the land on which he resides, owns or cultivates.
9 In Tujang Vaichong village, the villagers protested against the giving of *Changseo* or tributary gifts to the Chief of the village. So, the practice was stopped before my field visit.
10 Ibid.
11 Kaikhosei, S.L., Chief of Motbung, Interviewed on 13 October 2008 and S.L. Vumkhopao Lhouvum, Joint Secretary of Members of Village Authority.
12 It is the custom of the Thadou-Kukis in particular and the Kukis in general to be penalised for a wrong act in terms of pigs.
13 In 1954, Verrier Elwin was appointed advisor on Tribal Affairs to the North Eastern Frontier Agency. He became a citizen of India after Independence.
14 Manipur State Development Report, Planning Commission, Government of India, Institute for Human Development, New Delhi, http://manipur.nic.in/planning/DraftMSDR/Draft_SDR_pdf/Chapter%2017_Land%20Rights_Autonomy.Pdf, Accessed 10 July 2014.
15 Enoch Kipgen, Asst. Head Master, L.K. Junior High School, Tujang Vaichong Village, interviewed on 10 November 2009.

References

Anderson, Robert S. and Walter Huber (1988). *The Hour of the Fox: Tropical Forests, the World Bank, and Indigenous People in Central India*. New Delhi: Vistaar Publications.

Baviskar, Amita (1994). 'Fate of the Forest: Conservation and Tribal Rights'. *Economic and Political Weekly*, 29(38): 2493–501.

Bhowmick, P.K. (1980). *Some Aspects of Indian Anthropology*. Calcutta: Subarnarekha.

Das, J.N (1995). 'Customary Land System of Hill Ethnoses of Manipur', in Naorem Sanajaoba (ed.), *Manipur Past and Present: The Ordeals and Heritage of a Civilization* (Volume III/Nagas and Kuki-Chins). New Delhi: Mittal Publications.

Devi, P. Binodini (2006). *Tribal Land System of Manipur*. New Delhi: Akansha Publishing House.

Elwin, Verrier (1964). *The Tribal World of Verrier Elwin: An Autobiography*. London: Oxford University Press.

Gangte, Priyadarshni M., and Aheibam Koireng Singh (eds.) (2010). *Understanding Kuki since Primordial Times*. New Delhi: Maxford Books.

Haokip, T.T. (2009). 'Critically Assessing Kuki Land System in Manipur', In Ch. Priyoranjan Singh (ed.), *Tribalism and the Tragedy of the Commons: Land, Identity and Development, the Manipur Experience*, 304–28. Delhi: Akansha Publishing house.

Kipgen, Sheikhohao (2009). 'Land, Identity and Development: Perceptions Focusing on the Tribals of Manipur', In Ch. Priyoranjan Sindh (ed.), *Tribalism and the Tragedy of the Commons: Land, Identity and Development: The Manipur Experience*. New Delhi: Akansha Publishing House.

Kulkarni, Sharad (1987). 'Forest Legislations and Tribals Comments on Forest Policy Resolution', *Economic and Political Weekly*, 22(50): 2143–48.

Mandavdhare, S.M. (1993). 'Caste and Land Relations: Evidence from Marathwada', In Aijazuddin Ahmad (ed.), *Social Structure and Regional Development: A Social Geography Perspective*. Jaipur/New Delhi: Rawat Publications.

Munshi, Indra (2005). 'Scheduled Tribes Bill', *Economic and Political Weekly*, 40(41): 4406–9.

Ray, Asok Kumar (1990). *Authority and Legitimacy: A Study of the Thadou-Kukis in Manipur*. Delhi: Renaissance Publishing House.

Roy Burman, B.K. (1992). 'Historical Process in Respect of Communal Land System and Poverty Alleviation among the Tribals', In Chaudhuri, Buddhadeb (ed.), *Economy and Agrarian Issues*. New Delhi: Inter India Publications.

Savyasaachi (1991). 'A Study in the Sociology of Agriculture', Unpublished PhD thesis submitted to the University of Delhi, Department of Sociology, Delhi School of Economics, Delhi.

Savyasaachi (2001). 'Forest Dwellers and Tribals in India', In Susan Visvanathan (ed.), *Structure and Transformation: Theory and Society in India*. New Delhi: Oxford University Press.

Saxena, N.C. and Madhu Sarin (1999). 'The Western Ghats Forestry and Environment Project in Karnataka: A Preliminary Assessment', In Roger Jeffery and Nandini Sundar (ed.), *A New Moral Economy for India's Forests?* New Delhi: Sage Publications.

Sharma, Suresh (1994). *Tribal Identity and the Modern World*. New Delhi: Sage Publications.

Singsit, Seiboi (2010). *Traditional Forestry Management of the Thadou-Kukis*. New Dehi: Akansha Publishing House.

Sitlhou, Hoineilhing (2011a). 'Continuity and Change: The Setting of a Customary Court amongst the Kukis', In Ngamkhohao Haokip and Michael Lunminthang Haokip (ed.), *Kuki Society: Past, Present and Future*. New Delhi: Maxford Publication.

——— (2011b). 'Land and Identity: A Sociological Study of the Thadou Kukis of Manipur', Unpublished PhD thesis submitted to Jawaharlal Nehru University, Centre for Study of Social Systems, School of Social Sciences, New Delhi.

——— (2014). *Kuki Women*. New Delhi: Synergy Books India.

Sundar, Nandini, Roger, Jeffery (ed.) (1999). *A New Moral Economy for India's Forests?: Discourses of Community and Participation*. New Delhi: Sage Publications.

Tinker, George E. "Tink" (2008). *American Indian Liberation: A Theology of Sovereignty*. New York: Orbis Books.

Part II

Ethnicity, identity and belonging

Chapter 7

Identity, deprivation and demand for bifurcation of Meghalaya

Purusottam Nayak and Komol Singha

Introduction

Deprivation is a term used in social sciences to describe feelings or measures of economic, political or social disadvantage which is relative rather than absolute. Relative deprivation is the experience of being deprived of something to which one thinks he is entitled to. It refers to the discontent that people feel or perceive when they compare their positions to those of others similarly situated and find out that they have less than they deserve. It is a condition that is measured by comparing one group's situation to the situations of those who are more advantaged. Relative deprivation reflects a perception by a region/state/community that the circumstances or the lives of their people are not provided benefits to which they are justly entitled. When an ethnic group experiences relative deprivation, the potential for spontaneous outbreak of violence directed at rival groups intensifies (Hossain, 2009). It is also possible that a group might perceive or measure their condition wrongly without considering the other ends. The situation in the Northeastern Region (NER or region, hereafter) India including the state of Meghalaya is a bright example in this regard where decades of economic, political, environmental and social deprivations have forced the youth into militancy and violence. However, the demand for division of the state of Meghalaya in the recent past is not a clear cut case of deprivation but of intolerance among ethnic groups within the state. This chapter is nothing but an argument in this regard.

Meghalaya state

Meghalaya is one of the smallest states in the region, predominantly occupied by the three major tribes – Khasi, Garo and Jaintia. Each of them had their own kingdoms until they came under the British administration in the nineteenth century. However, other tribes, claimed to be the aboriginals of the state like Koch, Hajong, Rabha and Mikir are also living for years along with the major tribes. According to 2011 population census, the Khasi constituted around 45 per cent of the total population of the state, followed by the Garo with 32.5 per cent and the rest 22.5 per cent are from other communities including Bengali, Assamese, Nepali/ Gurkha and Hindi-speaking communities from the so-called mainland India. The state has a total geographical area of 22,429 sq km, and is surrounded in the east and north by the state of Assam and in the west and south by Bangladesh. In other words, the state is about 0.7 per cent of the country's total area and 8.6 per cent area of the northeastern region. Of the total geographical area, about 37 per cent is covered by the forest which is also notable for its biodiversity. Much of the forest is privately managed. The state government controls only area under the reserved forest, which is about 4 per cent of the total forest area. The climatic condition of the state, though varies with altitude, is moderate and humid. The state is also a storehouse of mineral resources. Some of the major minerals that are presently exploited are the coal, limestone, clay and sillimanite. Though the inhabitants of Khasi and Jaintia Hill districts speak a similar language, they have different dialects. The Garo Hill districts have very different customs and different languages. Though principal languages are Khasi and Garo, English is used as the official language in the state and they practice matrilineal system.

Ethnicity and state formation

As mentioned earlier, Meghalaya is the homeland of three major tribal communities – Khasi, Jaintia and Garo with their numerous divisions into clans. The term 'Khasi' is often used in generic sense and includes Khasi, Jaintia, Bhoi and War. They are collectively known as the 'Hynniewtrep' people and are mainly found in the four districts of east Meghalaya– namely East Khasi Hills, West Khasi Hills, Ri–Bhoi and Jaintia Hill districts. The Jaintias are also known as 'Pnars'. The Khasis occupying the northern lowlands and the foothills are generally called the 'Bhoi'. Those who live in the southern tracts are termed as the 'Wars'. In the Khasi Hills, the 'Lyngams' inhabit in the north-western part of the state. But all of them claim to have descended from the *ki-hynniew-trep* and are known by the generic name of Khasi–Pnars or simply 'Khasi' or *Hynniewtrep*. The Garos belonging to the Bodo family of the Tibeto-Burman race live in the western part of the state. They prefer to call themselves *Achiks* and the land they occupy as *Achik* land.[1]

Right after country's independence in 1947, when the All Party Hills Leaders' Conference (APHLC) of the then undivided Assam was formed, the leaders of the then North Cachar Hills (NC Hills) and Mikir Hills, too, joined it. As a result of which, the state of Meghalaya was created in 1972 following concerted efforts made by the combined leadership of the Khasis, Garos and Jaintias under the flagship of the APHLC (Upadhyaya et al., 2013). The people of NC Hills (presently Dima Hasao district) and Mikir Hills (presently Karbi Anglong district) who were living closely with the Khasis and the Garos decided not to join Meghalaya, though an option was given to them (Hussain, 1987; *The Assam Tribune*, 5 September 2013; Gohain, 2014).

Historically, under the Government of India Act 1935, the hill areas of undivided Assam were divided into two categories – One, the Lushai (Mizo) Hills and NC Hills which were classified as 'excluded area'. Two, the united Khasi and Jaintia hill districts with partial exception of Shillong town which was also the capital of Assam at that time, the Garo Hills, Naga Hills and Mikir (Karbi) Hills were classified as 'partially excluded area'. The Government of Assam had no jurisdiction over the excluded areas which were administered under the special power of the Governor. After independence, the Constitution also accepted broadly the spirit of the Government of India Act of 1935 by providing each hill district an Autonomous District Council with a fairly large autonomous power under the Sixth Schedule of the Constitution of India (Hussain, 1987). This led to the then hill districts of Assam, namely the Naga Hills, Khasi and Jaintia Hills, Garo Hills and Lushai Hills for the creation of new states one after another. The former Naga Hills district became the full-fledged state of Nagaland in 1962 and in 1972, the Khasi and Jaintia Hills and Garo Hills formed the state of Meghalaya and in the same year the Lushai Hills became a union territory and subsequently a full-fledged state of Mizoram in 1987. Other major part of the hill area – North East Frontier Agency which was under the control of Central Government of India and ruled through the state of Assam, became the union territory of Arunachal Pradesh in 1972.

Insurgency movement and conflict in Meghalaya

Although Meghalaya relatively is a peaceful state compared to some other states of the region, it has been riven by ethnic conflicts between the indigenous community and non-local immigrants since its formation in 1972. The steady rise of economic immigrants, mainly Bengalis from

Bangladesh, Nepalis from Nepal and other parts of India, resulted in uneasiness among the locals. The immigrants began to dominate business establishments, labour force and other employment opportunities. As a result, the state witnessed ethnic riots between indigenous tribals and immigrant non-tribal communities in 1979, 1987 and 1992, respectively (Haokip, 2013). Since the eighties numerous cycles of ethnic cleansing incidents rocked the state and people belonging to Nepali, Bengali, Bihari and Marwari communities became the target. In the 1990s, the Bengalis were the prime target of the ethnic violence. Since the early 1980s, an estimated 25,000–35,000 Bengalis have left Meghalaya to other parts of the country especially to West Bengal. In 1981, there were 119,571 Bengalis in Meghalaya, 8.13 per cent of the state's population. Ten years later in 1991, it was reduced to 5.97 per cent of population (Baruah, 2004; Phukan, 2013).

With the reclamation of tribal identity in the new state amid growing scarcity of resources led to a range of conflicts. The rise of ethnocentric politics emerged as the major plank around which much identity-based conflict transpired. Predicated on the cultural superiority of two tribal communities – the Khasis and the Garos over the non-tribal population, politically motivated ethnocentrism led to the commission of many dreadful acts against members of the non-tribal population. This trend was more conspicuous in the Khasi Hills, where the elevation of the Khasis to a dominant political position in the newly created state led them to challenge the hitherto ascendancy of the non-tribal population, who were often branded as 'Bangladeshis' – nationals of Bangladesh. Of the various causes of conflicts in Meghalaya, economic disparity emerged as the most prominent. The state's community-based agrarian economy lost much of its verve as a result of the unchecked privatisation of community land, while the decline of agrarian resources made it extremely difficult for members of the tribal population to maintain their livelihoods. As a result of which, the state first witnessed insurgent activities in the early 1980s and this took on a virulent aspect in the 1990s with the emergence of the Achik Liberation Matgrik Army (ALMA) and Achik National Volunteer Council (ANVC) in the Garo Hills, and the Hynniewtrep National Liberation Council (HLNC) in the Khasi Hills.[2] However, since the mid-1990s, there has been a relative change in the nature of ethnic relations between the communities. Although the relations between the tribals and the non-tribals relatively improved, ethnic tensions shifted to the so-called indigenous tribes in the recent past (Haokip, 2013).

Internal conflict and demand for bifurcation of Meghalaya

The spectre of unemployed youth haunts particularly Garo Hill region – the worst site of underdevelopment and poverty in the state. The situation in this area is all the more disconcerting for its inhabitants when they contrast their conditions to those in the relatively developed region of the Khasi Hills. This is what we refer to as deprivation. The Garo Hills' meagre infrastructure and essential services, scanty health and educational facilities and poor connectivity to the rest of Meghalaya accentuated the sense of relative deprivation in the region (Upadhyaya et al., 2013). The issue is – does the claim or demand of the community is recognised by the other group? If it is negated, often the end result is outbreak of conflict, or this is what we termed as *internal conflict* in NER.

The mushrooming of militant groups in Garo Hills becomes a cause of worry. While the ANVC and its splinter group, ANVC-B, are officially under ceasefire with the government, the Garo National Liberation Army (GNLA), the United Achik Liberation Army (UALA) and the Achik National Liberation Army (ANLA was formed in October 2013) are active in the interior areas of Garo Hills and in its adjoining areas of Assam and West Khasi Hills. Again, there is another group – the GNLA-F led by former GNLA militants Reading T Sangma, Jack Baichung and Savio R

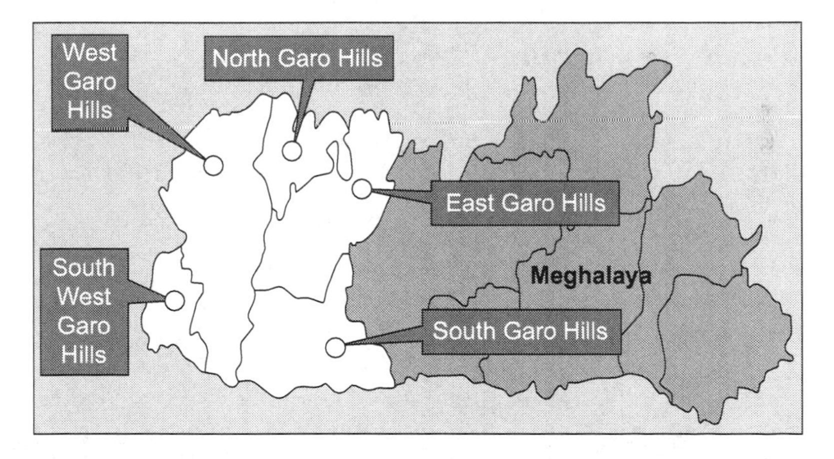

Figure 7.1 Separate state demanded by Garos in Meghalaya

Source: *The Telegraph*, Kolkata, Friday, 2 August 2013.

Marak. Meanwhile, ANVC suffered a further split in mid-November 2013 when seven members deserted the designated camp where they have been living since the ceasefire and formed a new group, adding to the murky scene.[3]

To be precise, the main bone of contention between the Khasis and the Garos was the implementation of the 1971 Reservation Bill. Secondly, the hegemony of the Khasis was felt during 2005 when the Meghalaya Board of School Education, which had its head office in Tura (in West Garo Hills), agreed to reorganise itself in Shillong (Khasi area) along the lines demanded by the Khasi Students Union. Undoubtedly, these episodes shaped the embittered situation between the Khasis and the Garos, which may develop into violence. The subsequent massive outcry resulted in demands by the Garos for a separate state (Upadhyaya et al., 2013). The demonstration and agitation have intensified and are being organised by the Garo Hills State Movement Committee, a conglomeration of various pressure groups and political groups. The demand for Garo is that the central government should consider the creation of separate Garo and Khasi–Jaintia states in Meghalaya based on linguistic lines as envisaged in the States Reorganisation Act, 1956. In short, the Garo National Council (GNC) and the Garo Students' Union (GSU) are demanding a separate state for the Garos on linguistic lines while the Hill State People's Democratic Party (HSPDP) is demanding a Khasi–Jaintia state on the other hand.

Argument of the study

Though it is a little happening, the state is facing internal conflict and it is getting worse day by day. Recently, as a part of its effort to regulate alienation of indigenous land, the state government proposed deleting certain Scheduled Tribes from the existing list of Scheduled Tribes in the state, leading to an agitation of the indigenous minorities living in the state (Haokip, 2013). The mute question is – who is being relatively deprived and by whom? If all the relatively smaller communities start demanding separate states or taking up arms for their discontent that they feel or perceive when they compare their positions to others, there will not be an end to further division of the states in the country and armed conflict in the region. It is a reign of terror especially in the region now. It is much hated by everyone in the society but few dare to speak up and for those who speak up would be bashed up. Also, there is no guarantee that the present demand for bifurcation of Meghalaya will not have further reorganisation or bifurcation of it in the near future.

Demographic equation is concerned; Meghalaya cannot be compared with Mizoram or Nagaland. It is because, Shillong, the capital of Meghalaya, was a British outpost and later the capital of the then undivided Assam. Therefore, Shillong had a sizeable non-tribal population and the same holds true for the large sections of Garo Hills particularly areas bordering Goalpara in Assam. The people of Meghalaya cannot just forget those histories and pretend to write on a clean slate (Mukhim, 2013). Violence against the outsiders or others is not the panacea for all. The insiders (aboriginals) become outsiders when they move out. If so happens, the same method of violence can be applied to them. People of Meghalaya or NER should not move out of their respective habitats or homelands as they defined. One should consider the limitation of the others and should respect other communities as well. In this context, seeing the new ethnic movements in Meghalaya, Mukhim (2013) raised few questions – when we need a good doctor do we check his tribe, caste, class or do we repose our faith on his/her expertise and credentials? The same is the case with a good teacher or lawyer. So if we are interdependent then is it not fair to share a slice of the cake with those who strive to build Meghalaya as much as the tribals do?

As discussed earlier, the Garos belongs to the Bodo family who are concentrated in present Assam and spearheading separatist movement for theirs. As the demand for territorial integration and bifurcation of the states based on the linguistic, ethnic or geographical lines has been the order of the day in the NER, there is no guarantee that Garos will not demand for integration with Bodos of Assam. On the other hand, there is no guarantee that the Jaintias, Bhois, Wars or any other minority tribal communities of Meghalaya who have already fabricated in the larger Khasi identity will not demand for separation from the larger confederation. Had the Karbis and Mikirs of present-day Assam joined Meghalaya when they were given an option to merge with Meghalaya (as they revolted against Ahoms together), there could have been different politics of the present Meghalaya.

Conclusion

As we have discussed, though Meghalaya is relatively better off compared to sister states in NER in terms of conflict and violence, one cannot completely ignore the tension between the indigenous tribal communities and immigrants since state formation in 1972. The state has experienced a series of communal riots since 1980s. However, in the recent past, the tensions in the state have shifted towards the internal feuds among the indigenous tribes, especially Khasi and Garo. Given that economic insecurity especially employment opportunities in the public sector is the primary cause of tension between the majority indigenous communities in the state, equitable distribution of economic resources would be one giant step towards reducing discontentment in the backward areas (Haokip, 2013). Therefore, politicising and polarising the communities for personal vote stunt opportunities will exacerbate the present crisis. One should understand the root cause of the problem, not the consequence alone.

Also, efforts to achieve peace through the appeasement policy of economic incentives or peace talks are unlikely to endure any sustainable settlement in the state. Therefore, commitment while negotiating conflict from the government and community leaders should be maintained. Also, one should not forget that the Constitution of India guarantees certain fundamental rights to all citizens irrespective of their caste/creed/tribe. The tribes of the region have been enjoying special protection under the Sixth Schedule of the Constitution. But other communities in the region do not enjoy such protections. Should we call them as relatively deprived section? Therefore, it is good to learn from other countries where a shared citizenship based on affinities rather than differences defines human relationships. The more we stress on differences the sharper will be the contours for conflict. As the people/community in any state of the region is the confederation of many tribes and communities, we need to learn to live together and solve the problems together. Since the economic factor is identified as a major bone of contention between the two major indigenous communities in Meghalaya (Khasi and Garo), policymakers should emphasise more on the equal distribution of development initiatives in the state. One should also remember that the bifurcation of the existing state is not the panacea for the present ethnic crisis. Equitable distribution of resources and reduction of inter-district disparity should be given more emphasis.

Notes

1 Karna, M.N. http://www.ide.go.jp/English/Publish/Download/Jrp/pdf/133_7.pdf.
2 See Upadhyaya et al. (2013).
3 See Das (2013).

References

The Assam Tribune (2013). 'New Forum to Espouse Separate State Demand', *The Assam Tribune*, Thursday, 5 September 2013, accessed 14 March 2014: http://www.assamtribune.com/scripts/detailsnew.asp?id=sep0513/state06.

Baruah, Apurba K. (2004). 'Ethnic Conflicts and Traditional Self-Governing Institutions: A Study of Laitumkhrah Dorbar', Working Paper No. 39, Crisis States Programme, Development Research Centre, Destin, LSE, Houghton Street, London.

Das, Rani P. (2013). 'Assam and Meghalaya: Threats of Violence in Garo Heartland', Article No. 4192, 26 November 2013. Institute of Peace and Conflict Studies, Guwahati.

Gohain, Hiren (2014). 'A Note on Recent Ethnic Violence in Assam', *Economic and Political Weekly*, 49(13): 19–22.

Hossain, Md. Kamal (2009). 'Fraternal Relative Deprivation of a Tribal Population in Chittagong Hill Tracts, Bangladesh', *Journal of Life Earth Science*, 3(4): 29–32.

Hussain, Monirul (1987). 'Tribal Movement for Autonomous State in Assam', *Economic and Political Weekly*, 22(32): 1329–32.

Haokip, Thongkholal (2013). 'Inter-Ethnic Relations in Meghalaya', *Asian Ethnicity*, accessed 8 April 2014. DOI: 10.1080/14631369.2013.853545.

Karna, M.N. (Undated). 'Identity Politics', accessed 14 March 2014: http://www.ide.go.jp/English/Publish/Download/Jrp/pdf/133_7.pdf.

Mukhim, Patricia (2013). 'Non-Tribals in Meghalaya – Non-Citizens or Half Citizens?' *The Shillong Times*, Friday, 22 March.

Phukan, Mridula Dhekial (2013). 'Ethnicity, Conflict and Population Displacement in Northeast India', *Asian Journal of Humanities and Social Sciences*, 1(2): 91–101.

Upadhyaya, Anjoo Sharan, Priyankar Upadhyaya and Ajay Kumar Yadav (2013). 'Interrogating Peace in Meghalaya', Core Policy Brief No. 03. Cultures of Governance and Conflict Resolution in Europe and India, Peace Research Institute, Oslo.

Chapter 8

Identity politics, conflict and development among the Mizos in Mizoram

Lalrintluanga

Introduction

The concept of identity may be understood as 'the sameness of a person or thing at all times in all circumstances; the condition or fact that a person or thing is itself and not something else' (Simpson and Weiner, 1989: 620). Ethnic identity is, thus, an affiliative construct, where an individual is viewed by themselves and by others as belonging to a particular ethnic or cultural group. Therefore, an individual's association or identification with a group has always been influenced by racial, natal, symbolic and cultural factors cemented by his/her 'psychological attachment to an ethnic group or heritage' (Cheung, 1993: 1216). One of the psychologists Jean Phinney (2003: 63) also maintains that, 'ethnic identity refers to one's identity or sense of self as a member of an ethnic group'. From her perspective, one usually claims an identity is based on a common ancestry and shares at least a similar culture, race, religion, language, kinship or place of origin. Ethnicity, in this context, stands for a group's way of conceptualising and relating to society and thus welds together individuals who share a history, culture and community, who have an amalgam of language, religion and regional belonging in common and they feel that they come from the same stock (Wilson and Bodil, 1991: 2). Individuals may rely on labels to describe their ethnic affiliation and subsequently their identity. It is with the help of ethnic labelling that ethnic identity of a person can be exposed to others.

Identity always has political implications thereby giving rise to identity politics on ethnic lines. In most cases, identity politics have been accompanied by ethnic conflicts arising out of 'an active disagreement between people with opposing opinions or principle' (Woodford and Jackson, 2003: 620). Usually, conflicts occur when two or more individuals or ethnic groups feel that they have irreconcilable differences, or when they feel that their needs, interests or wants are in jeopardy at the hands of other groups. All over the world, there are different ethnic groups having different perspectives on life and hence clashes of interest become inevitable. When clashes of interest arise between two or more ethnic groups, we call it ethnic conflict. In most cases, conflicts have resulted from a disagreement through which the parties involved perceive a threat to their needs, interests or concerns. In this connection, according to Lewis (1956: 3), 'struggle over values and claims to scarce status, power, and resources in which the aims of the opponents are to neutralise, injure, or eliminate their rivals'.

'Development' is a multidimensional concept which may be social, cultural, economic or political. However, our primary concern in the present context is nothing else but economic development of a country or region for ensuring human development as related to standard of living and quality of life. In this sense, economic development is concerned with economic wealth of a country or a particular region for the well-being of its inhabitants at each level. Professor Dudley Seers argues that development occurs with the reduction and elimination of poverty, inequality and unemployment within a growing economy. In support of Seers' argument, Professor Michael Todaro (1977: 12) also sees three objectives of development:

- Producing more 'life-sustaining' necessities such as food, shelter and health care and broadening their distribution.
- Raising standards of living and individual self-esteem.
- Expanding economic and social choice and reducing fear.

From a policy perspective, economic development can be understood as efforts that seek to improve the economic well-being and quality of life for a community by creating job opportunities and supporting or growing incomes and the tax bases. Therefore, the economic administration must make sure that there are sufficient economic development programmes in place to assist the people to achieve their goals. Those types of programmes are usually policy created and can be local, regional, state-wide and national in nature.

In the case of the indigenous tribes and subtribes of Mizoram, such as Lusei, Ralte, Hmar, Paite, Pawi and those minor tribes namely Khawlhring, Khiangte, Chawngthu, Chawhte, Ngente, Renthlei, Tlau, Pautu, Rawite, Zawngte, Vangchhia, Punte and so on, identity politics, conflicts and development are deeply rooted in their minds. At the beginning of their inclusion as the people of Assam, some of the ambitious elite and political leaders of these indigenous Mizo tribes and subtribes could successfully bring the ethnic minority groups together – based on their shared common origin, common customs and traditions and linguistic affinity – to build a common ethnic identity against the possible erosion of their cultures and identity by the plains people. When the Mizo National Front (MNF) came into existence to protest against the lukewarm attitude of the Assam Government towards the Mizo District during *Mautam* famine which broke out in 1959, their avowed objective was establishment of 'Independent Mizoram' comprising all the Mizo-inhabited areas (MNF Memorandum, 1965). Excepting one or two tribes, all other Mizo tribes such as Hmars, Paites, Pawis, Kukis, Zomis and Gangte, who were yearning for integration of the Mizos and their inhabited areas, had joined this new party. The MNF took up arms to fight for independence and this armed struggle continued up to the next 25 years. It is, therefore, evident that common identity awareness among the Mizo tribes was consolidated by the MNF's movement for independence which, as one of its political demands, included 'Greater Mizoram' for all the Mizo tribes living within and outside Mizoram.

A reminiscence of events prior to the MNF's movement for independence has crystallised that the Mizo Union (MU), the first political party formed in 1946, had already demanded the unification of all Mizo-inhabited areas in Assam, Manipur and Tripura into a single administrative unit which would be called 'Mizoram State'. This MU's integrationist movement had successfully attracted the sentiments of the Mizo tribes living in other areas of Northeast India which slowly diluted clan's politics spearheaded by some of the Mizo tribes living outside Mizoram. As the MU was advocating integration of all the tribes of Mizo living within and outside Mizoram, most of the Hmars in Manipur, Cachar and Mizoram had joined the MU. Eventually, Hmar People's Conference (HPC) had joined the MU's struggle for Statehood of Mizoram (Das, 2007: 41). It will not be an exaggeration to point out that, in those days, there was no serious issue of separatism for the Hmars who 'had full confidence in the leadership of the Mizo Union of Mizoram' (Lalthanpari, 2014: 362). Various units of MU were also set up without any difficulty in the Mizo-inhabited areas of Assam and Manipur and some of the members had occupied leadership positions in the party (among them were Pachhunga and H.K. Bawichhuaka from Hmar community). In those days, the Hmar and the Paite communities were having a high hope that the formation of the Autonomous Mizo District would bring about integration of their inhabited hill areas of Manipur, Cachar and North Cachar Hills of Assam to form one administrative unit. Unfortunately, when the formation of an Autonomous

Mizo District took place in 1951, the MU's political demand for integration of all the Mizo-inhabited areas was nowhere in the picture. In those days, there was a popular conjecture among the Mizo public that the State Government had hurriedly recommended the formation of the Mizo District with an ulterior motive of subsiding the MU's strong demand for Mizoram Statehood. In the absence of any specific provision in the Sixth Schedule for integration of all the Mizo-inhabited areas into one administrative unit, in the 1960s, the Hmars had raised their voices for 'consolidation of the Hmar areas of Manipur and the Mizo and North Cachar Hills' (Chaube, 1973: 210–11).

After several rounds of peace negotiations, the underground MNF leaders and the Government of India could resolve their differences to put an end to the 20-year-old Mizo insurgency and ultimately, the MNF supremo, Laldenga and the Government of India had signed the Memorandum of Understanding (MoU), popularly known as 'Mizo Peace Accord', on 30 June 1986. Apart from ending the MNF's armed struggle for independence, this 'Peace Accord' had led to the elevation of the Union Territory of Mizoram to a full-fledged Statehood on 20 February 1987 but 'left out the demand for a Greater Mizoram that would have integrated Hmar areas in Manipur into Mizoram' (Rajagopalan, 2008: 25). It was, therefore, very unfortunate for all the Mizo ethnic tribes permanently living in Assam, Manipur and Cachar areas that the 'Peace Accord' did not contain any provision for the materialisation of their long-awaited integration of all the Mizo-inhabited areas to form 'Greater Mizoram'. For the past 20 years of Mizo insurgency, it was the aspiration of most of the Mizo ethnic tribes that the signing of 'Mizo Peace Accord' by the MNF leaders and the Government of India would facilitate the formation of 'Greater Mizoram' where they could find an opportunity to enjoy peace and prosperity. As their dream for 'Greater Mizoram' did not come true, the new state of Mizoram had to inherit the same boundary of the erstwhile union territory of Mizoram (Neitham, 2011).

It will not be an exaggeration to state that, prior to the signing of the 'Mizo Peace Accord', the process of formation of ethnic identity among the Mizo tribes was based on the idea of larger group formation. But, the failure of the 'Peace Accord' to incorporate a provision for integration of the age-old Mizo-inhabited areas to form 'Greater Mizoram' had prepared a fertile ground for small group identity assertion in the Mizo society. Out of desperation, some of the Mizo ethnic tribes had openly begun to assert themselves as distinct from other ethnic groups. In this connection, sociologist M.N. Karna, also rightly said, 'Several smaller groups, till now forming part of the larger Mizo grouping, have started asserting their separate identity by calling themselves by different names such as Zomi, Zou, Mara, Hmar, etc.' (Karna, 1998: 229–30). Hence, identity assertion based on smaller ethnic groups had begun to unleash internal conflicts in their simple, close-knit and gregarious Mizo society. One such example of internal ethnic conflicts experienced by the Mizos is the recent ethnicity-based autonomy movements spearheaded by the Hmar People's Convention (HPC-D). How good the intention of ethnic assertion by some ethnic tribes might be, these movements have always been accompanied by insurgent activities which, more often than not, have adverse effects on vital process of economic development particularly inside their demand area for setting up of a separate Council for the 'Hmars'.

It is pertinent for one to note that, when interstate boundaries were drawn by the Government of India immediately after independence, some parts of the age-old Mizo-inhabited areas were sliced out to form parts of other northeastern states contiguous to the present Mizoram. Despite the fact that the Mizo ethnic tribes are the original settlers of the aforesaid northeastern states, namely Assam, Manipur and Tripura, they, as ethnic minorities, feel deprived of any opportunity for socio-economic development in their respective States. Truly speaking, it was their urge for attainment of socio-economic development that compelled some of the Mizo ethnic tribes to lend support to the demand for 'Greater Mizoram' first raised by the MU – the first political party that the Mizos had ever formed – and secondly by the MNF. However, experiences have shown that ethnic conflicts do not pay but derail the ongoing socio-economic development in general and vital development indicators in the areas where ethnic conflicts have taken place.

Profile of Mizoram and its indigenous inhabitants

Mizoram is one of the eight states of northeastern region of India and was formerly known by different names. It was first known by the Lushai Hills and, later on, by the Mizo Hills of the composite state of Assam. The geographical area of Mizoram is 21,087 sq km (Government of Mizoram, 2000: viii). It is a mountainous and hilly state lying in the extreme southern corner of the region; it consists of six parallel hill ranges which run mostly from north to south with a tendency to be higher in the east of the territory and tapering in the north and south 'enclosing between them deep river valleys' (Thanga, 1978: xii). The southern mountain ranges join Arakan Yoma of Burma (Myanmar). In the absence of proper border management between India and Myanmar till the recent past, insurgents could easily find sanctuary on either sides of this international border and thus, could have easy access to India to cause threat to our national sovereignty. Mizoram is bounded on the east by Burma and on the west by Bangladesh. It is also bounded in the north by Manipur and Cachar district of Assam and in the north-west by Tripura.

Mizoram has a long international boundary with its neighbouring countries, which is not well guarded by the concerned law-enforcing authorities. Three quarters of its boundary are international having a common open border with Bangladesh over a length of about 318 km (Government of Mizoram, 2000: viii) and with Myanmar another 404 km (Government of Mizoram, 2000: viii). The indigenous people residing in Mizoram are from the different Mizo tribes who are believed to have migrated to their present habitat from the east. However, it is difficult to exact the year when these different tribes began to call themselves 'Mizos'. But, most of the local historians have believed with unanimity that the people had begun to assume the term 'Mizo' for calling themselves in about 1765 AD when they were living at a small township called 'Zopui' in the western part of Burma (Myanmar) adjacent to the present eastern boundary of Mizoram. In short, various tribes and clans inhabiting the present Mizoram and abroad who are knit together by common language, common customs and common traditions have called themselves 'Mizo' and their land has always been referred to by the proper Mizos as 'Mizoram'. Barkataki (1964: 82) has also rightly pointed out this fact.

In spite of the adoption of a collective name 'Mizo' for calling themselves, they were vaguely given different names, at different points of time, by non-Mizo writers. In Burma, they were called 'Chins' by the Burmese which literally means 'man with the basket' (Lalthangliana, 1975: 69). The earliest Mizos who migrated to India from Burma were known as 'Kukis' and the second batch of Mizo immigrants were called 'New Kukis'. The last group of the Mizo tribes who migrated to India were 'Luseis'. In Manipur, they were also known as 'Kukis' which means 'wild hill tribe'. Similarly, they were called 'Kukis' by the Bengalis who lived in Cachar, Tripura and Chittagong areas adjacent to Mizoram boundary (Lewin, 1986: 98). During the time of Warren Hastings, the British rulers were said to have followed the Bengalis in referring to the Mizos as 'Kookis' or Kuki (Mackenzie, 1884: 437). The name 'Kukis' continued to be applied to the whole group until 1871 (Mizo Union Memorandum, 1947).

The 'Mizo' as an ethnic group of Mongoloid stock (Thanga, 1978: 3) has been formed as a result of the assimilation of many original hill tribes, subtribes and clans. According to one Mizo historian, Rev. Liangkhaia, those original hill tribes include Lusei, Hmar, Ralte, Paite, Pawi (now 'Lai'), Lakher (now 'Mara'), Pang and their allied tribes (Liangkhaia, 1976: 19–21). The transformation of these hill tribes and clans into a common ethnic identity is said to be brought about by the deliberate 'move for broad basing of the ethnic identity so as to involve all the tribes living in an area in the struggle for certain basic interests rather than rely on a few hundred or a few thousand people belonging to a single tribe' (Sharma, 1985: 163–4).

It is relevant to narrate that, when the Mizo tribes moved out of Burma, they 'lived together family by family and each sub-tribe settled in separate villages and had no Chief' (Thanga, 1978: 7). Inter-tribal feuds were a recurrent feature among themselves. So, in order to fight against frequent raids from more powerful Burmese from the east, a number of subtribes had felt the need of combining themselves together to form one big village which they built a Selesih in the east in 1740 (Zawla, 1976: 14). Selesih was a large settlement and was something like a confederation where several major tribes and subtribes of the Mizo origin, such as Lusei, Ralte, Hmar, Paite, Pawi and those minor tribes namely Khawlhring, Khiangte, Chawngthu, Chawhte, Ngente, Renthlei, Tlau, Pautu, Rawihte, Zawngte, Vangchhia and Punte, lived together without any feuds among themselves. It was during their stay at Selesih township that a good beginning was made by different ethnic tribes for Mizo sociopolitical solidarity. Since that time onwards, majority of these tribes and subtribes have identified themselves as Mizos and have accepted, without any reservation, the term 'Mizo' as a single collective name to mean all those original hill tribes of the Assam-Burman subgroup that branches from Tibeto-Burman group of the main Tibeto-Chinese race (Lalthangliana, 1975: 2). George T. Haokip is very right in saying that, while various groups of Naga tribes have been generically called as Nagas, the Lushai and other allied tribes of the Mizo hill areas call themselves by a common name, Mizo (Haokip, 2012: 223–4). Though the British authority introduced 'Lushai Hills' as a new name for Mizoram whose literal English translation is land of the Mizos, the indigenous Mizo ethnic tribes never accepted it because there is no such Mizo word 'Lushai' and this word could be a wrong spelling of 'Lusei', one of the tribes constituting 'Mizo'.

When the British made an inroad into the Mizo-inhabited areas, they found that the Luseis (but not 'Lushai') were predominant among the tribes constituting 'Mizo' and their spoken dialect known as '*Duhlian* dialect' has developed into the major language spoken by such major tribes as the Hmar, the Paite, the Pawi and the Ralte living in Mizoram though these tribes originally had their own dialects (Shakespear, 1912: 40). Some of the tribes of Mizo origin like Pawi, Lakher, Paite, Hmar living in southern and northeastern corners of and outside Mizoram can still sustain their separate dialects and cultural identities. However, the language which is now commonly understood, used and spoken as a *lingua franca* by different tribes and clans of the Mizo tribes is 'Mizo' (Liangkhaia, 1976: 90). In the meanwhile, quite a few number of Mizo tribes ever consider Mizo language as the Lusei dialect. If any scholar makes analytical study of the Mizo language without prejudice, he or she must find out that, with the incorporation of many words from the dialects of other Mizo tribes into it, the Lusei dialect has been transformed into the common language for all the tribes and clans of the Mizo people (Assam Secretariat, 1890: No. 1–46).

With regard to their social life, the Mizos maintained a simple, close-knit and gregarious society where there was no caste system (Soppit, 1898: 10). Till the signing of the Mizo Peace Accord in 1986, the feeling of clan distinction was virtually non-existent among the Mizo tribes in Mizoram. Hassan (2006: 18) has rightly explained the significant reason responsible for this absence of clan distinction among the Mizos and said, 'the term "Mizo" lacks any single ethnic marker, making it possible to integrate both Lushai and non-Lushai speaking communities into one united political voice'. In the meanwhile, some tribes of Mizo origin living outside Mizoram who are directly or indirectly influenced by other cultures still have a problem of tribe or clan feeling and prefer to be called by their tribe's or clan's names. In spite of this sign of withdrawal from the larger Mizo society, the fact remains that, on many occasions, some of them readily accept 'Mizo' as their nomenclature when the term 'Mizo' brings in some advantages; and one of such advantages is Tribal Scholarship for any Mizo tribes. With the passage of time, the process of economic development has given way to greater emphasis on monetary value and the consequent mad race in making easy money, which, according to Chaube, 'has corrupted much of the unsophisticated attitudes of the hill men' (Chaube, 1973: 107). In fact, it is this materialist impulse which has driven some of the educated elites among other Mizo tribes to go for small ethnic group identity assertion so that they could easily elevate themselves to power positions at the cost of the interest of the whole Mizo society.

If one makes a serious attempt to find out the reasons responsible for the declining social relations among different Mizo tribes, he/she will find that the urge for material prosperity and higher sociopolitical status which are very difficult to achieve in the larger Mizo society has tempted those playing leadership role in smaller ethnic groups to 'articulate their demands for increased share in political power, more opportunities for economic well-being and protection of their culture and tradition' (Karna, 1998: 229–30). So, the most effective instrument at their disposal for sustaining their demand is identity politics based on ethnic lines.

Development indicators in Mizoram

There may be many development indicators. However, development indicators in one country or state may not necessarily be relevant for other countries or states. Even among the Mizos living within and outside Mizoram, the most significant development indicators under their old economy were (a) self-sufficiency of a family in food through shifting jhum cultivation, (b) a family with a large number of members as workforce to cultivate the hard lands and (c) a family having a large number of domestic animals. But, the Mizos' perception of development indicators had changed after the introduction of democratic institutions of Autonomous District and Regional Councils in their ancestral land which, by that time, formed a part of Assam state. However, the Mizos' mindset in relation to development indicators had drastically changed after the elevation of their Autonomous District to a Union Territory status on 21 January 1971. Unlike before, this new constitutional status had given the Mizos an advantage to live on the free flow of development funds, in terms of crore, from the Central Government. Within a short spell of time, large-scale developments which had taken place in Mizoram have changed the Mizos' mindset with regards to development indicators. As of today, there is no big difference between the Mizos' perceptions vis-à-vis their fellow citizens on development indicators. Some of the relevant development indicators in Mizoram are as follows:

(a) **Communication network**: Good road communication network is recognised as one of the most indispensable infrastructures for economic development in Mizoram. Hence, availability of adequate infrastructure, like road communication facilities, acts as the precondition for sustainable social and economic development of the Mizo society. Unlike in other parts of the country, Mizo's aspiration to have good road networks had led to the construction of Aizawl-Lunglei Jeep Road by the people on a voluntary and self-help basis (Government of

Mizoram, 1990: 67). Category-wise distribution of the current road network in Mizoram is given in Table 8.1.

The 6,840 km length of road covers many villages and towns in Mizoram. To be more precise, out of the total 764 villages in Mizoram, 23 towns and 341 villages have been connected by all-weather roads. Besides, 338 villages are connected with fair weather roads whereas 85 villages remain unconnected by any type of roads (Records of Public Works Department, Government of Mizoram).

Of late, Mizoram bas been connected with the rest of the country by railways (though it is not fully operational) and air communication networks. The new airport of the state is located in Lengpui village, 44 km west of Aizawl. This Lengpui airport becomes the second largest airport in the entire region. Even then, development of transport infrastructure remains the prerequisite for economic development of the state. Hence, the low quality and limited accessibility or disparity of transport infrastructures can adversely affect the prospect of economic development in Mizoram. The crying need of the people to carry on their economic activities during rainy season is good quality of road.

(b) **Energy requirement and per capita electricity consumption**: Mizoram has rich power potentials which are not yet fully exploited for the development of the state. According to the Experts' estimate, there is 2,400 MW of hydro-power potential in Mizoram (NER Vision, 2020: 23). Unfortunately, the State Governments of yesterday and today have not yet harnessed even 10 per cent of the total power potentials in Mizoram. The hydroelectric projects that have been undertaken for execution are given in Table 8.2.

Table 8.1 Position of road communication network in Mizoram

Category of roads	*Length in km*	*% of total length*
National Highways (NH)	885	12.94
State Highways (SH)	225	3.29
Major/Other District Roads (M/ODR)	3,471	50.75
Village roads	935	13.67
Road's within towns and villages	704	10.29
Other purpose link roads	620	9.06
Total	**6,840**	**100.00**

Source: Records of Public Works Department, Government of Mizoram.

Table 8.2 Hydroelectric projects under construction in Mizoram

Sl. No	*Name of the project*	*Installed capacity (MW)*
1	Tuirial HE Project	60
2	Tuivai HE Project	210
3	Bairabi HE Project	80
4	Kolodyne-Stage I and II	500

Source: Government of Mizoram, 2010, 19.

When all these are executed by the State Government, they would be able to supply more than 10 per cent of the estimated potentials in the state even during the period of the leanest water discharge (Centre for North East Studies and Policy Research, NER Vision 2020: 23). In spite of this unsatisfactory exploration of power potentials available in the state, the energy requirement and rate of consumption for the people are increasing day by day. Mizoram requires 120 MW of power during summer, against a supply of 80 MW by the thermal and mini hydel projects of the North Eastern Electric Power Corporation and National Thermal Power Corporation in the region. If one makes a reality check, he/she will find that, out of its current peak power demand of 120 MW, the state can only manage to get 55–60 MW each day (*The Telegraph*, Calcutta, 2012). This indicates that, since the signing of the Peace Accord, none of the Governments which have ever been formed in the state has ever succeeded in generation of power energy. With the exception of the people's conference, most of the political parties in Mizoram have not yet formulated any clear-cut policy with political commitment in power sector. Table 8.3 shows the increasing rate of power consumption of the people in Mizoram for more than two decades.

In spite of the slow process of development in power sector, the growth rate of per capita electricity consumption of the people during the last 29 years in Mizoram has been 15.6 per cent. Of all the states, Mizoram has been placed on the top with per capita electricity consumption of 377 kWh in 2009–10. Electricity is an essential commodity for promoting living standards and the level of material progress of a country, and the very living standard or development is often measured in terms of per capita electricity consumption.

(c) **Economy**: The Mizo economy is an agricultural economy and 'the most important occupation of the society to sustain their lives was through jhuming' (Chatterji, 1975: 9). Till the early 1970s, majority of them were engaged in agricultural activities to sustain their lives. The

Table 8.3 Increasing rate of power consumption in Mizoram (in MW)

1981–82	*1989–90*	*1999–2000*	*2009–10*	*Growth rate*
5.6	65.0	120.8	377.0	15.6

Source: Singha, 2012, 408.

modus operandi was what is called 'shifting cultivation' on 'slash-and-burn' method of cultivation' (ibid.: 9). While it was extremely difficult for one to amass wealth over and above others under the old economy, the British rulers had introduced money economy which required the people to earn in cash or convert their daily agricultural products into cash 'to meet the Government annual house tax of three shillings' (McCall, 1949: 177). Thus, the earlier mode of evaluation of economic wealth among the Mizos was shifted to that of earning in cash, which gave material advantages over those who earned their living by cultivating the hard lands. Though Mizoram has good potentials for economic growth in the areas of agriculture, horticulture, tourism, handicraft and other areas, no big achievement has ever been made in these areas due to poor economic infrastructure.

(d) **Industrialisation:** Industrialisation can play a pivotal role in development processes of any region or country. Development of industries can effectively increase income, output and employment and accelerate the rate of growth of a backward area. Therefore, industries tend to exercise profound influence on other sectors of the economy including agriculture (Misra and Puri, 2007: 478). While such is the case, industrialisation in Mizoram is still at a rudimentary stage and the contribution of industrial sector in the state's economy is negligible. Due to the absence of large and medium industries, the entire Mizoram has been notified as backward and is categorised as 'No Industry State' (Government of Mizoram, 2008: 51). Hence, the contribution of industry, both registered and unregistered, hardly reaches 1 per cent to the GSDP (Government of Mizoram, 2008: 51). Small-scale and cottage industries are best suited in Mizoram as they require less capital, low levels of technical skills and less managerial experience. Handloom and handicraft industries play a prominent role in the socio-economic development of the state due to the possibility of exploiting the vast natural resources and local raw materials more efficiently by setting up small industries.

(e) **Human resource development:** In Mizoram, formal education was introduced for developing their required human resources by the

Christian Missionaries. After independence, the period 1950–70 saw a rapid increase in the number of educational institutions – primary to high schools – in Mizoram partly due to the active role played by the Government in the field of education. Table 8.4 shows the progress of literacy in Mizoram.

Since 1970, there has been a very rapid increase in the number of educational institutions from primary schools to colleges due to the leading role played by the Government of Mizoram in the field of education. As a result, Mizoram could become one of the states of India making best performance in human resource development.

(f) **Per capita income and domestic product:** The living condition of the people is always determined by the economic development of the country or state in which they live. So, the living condition of the people in a particular country or state can be conveniently measured in terms of the per capita income and domestic products of its people. Similarly, in order to understand the living condition of the people in Mizoram or its particular area, one has to look into the level of economic development which has ever been attained by the people in terms of the per capita income and domestic products within a particular period of time. Table 8.5 shows the per capita income and domestic products of Mizoram.

There may be various factors contributing towards the improvement of the per capita income of a country or state. However, what

Table 8.4 Progress of literacy rate in Mizoram

Year	*Total population*	*Percentage to total population*
1901	82,434	0.93
1911	91,204	3.98
1921	98,406	6.28
1931	124,404	10.71
1941	153,786	19.48
1951	196,202	31.13
1961	226,063	44.00
1971	332,390	53.79
1981	493,757	59.88
1991	698,756	82.27*
2001	888,573	88.80**

Source: Directorate of Economics and Statistics, Mizoram (1981).

*Government of India, *Census 1991*, Mizoram (Directorate of Census Operation, Mizoram), p. 76.

**Government of India, Census 2001, Mizoram (Directorate of Census Operation, Mizoram), p. ii.

Table 8.5 NSDP and PCI of Mizoram (at constant prices 2004–05)

Years	*NSDP (Rs in Lakh)*	*PCI (in Rs)*
2004–05	239,960	24,662
2005–06	257,728	25,826
2006–07	269,272	26,308
2007–08	298,845	28,467
2008–09 (P)	341,380	31,706
2009–10 (Q)	390,080	35,323

P: Provisional Estimates; Q: Quick Estimates.
Source: Government of Mizoram (2010), p. 36.

a country or state needs to increase its per capita income are quality manpower resources, peace and tranquillity. In fact, no investment of different kinds for increasing the per capita income will be possible in an area where multiple instances of lawlessness like armed revolt, looting and extortion cases are experienced by the people. Unless all these necessary preconditions for capital investment are fulfilled, no state or region can effectively set investment process on the right tract in order to increase its per capita income. It is, therefore, evident that the people in Mizoram can expect an increase in their par capita income only when a variety of factors collectively play their roles without restraints.

Impact of identity politics on development in Mizoram

Identity politics had emerged among the Mizo ethnic tribes immediately after the elevation of Mizoram to a full-fledged state mainly due to the failure of the 'Mizo Peace Accord' to contain a clause for the materialisation of 'Greater Mizoram'. Apart from other ethnic tribes of Mizo origin living outside Mizoram, the Hmars felt that they were let down by the MNF who previously preached 'Greater Mizoram' for the integration of all the Mizo-inhabited areas. Out of frustration, on 4 July 1986, the Hmars had disassociated themselves from the Mizos to launch a separate political movement from the platform of Hmar People's Convention (HPC), an incarnation of Mizoram Hmar Association (MHA). Thus, the HPC came into existence as one of the first ethnicity-based political parties of the post-Accord period in Mizoram, which declared its avowed objective to fight for the demands of the Hmar people living in the state of Mizoram. It is, however, relevant to mention that the HPC's initial demand was for the creation of *Hmar ram* (Hmar land; *First*

Memorandum of HPC, 1987) in the Hmar-concentrated areas of the region, namely Aizawl District of Mizoram, Cachar and North Cachar (NC) Hills of Assam, Churachandpur District of Manipur and adjoining Hills of Tripura (Lalthanpari, 2014: 363). As the original demand of the HPC was likely to affect the interests of the residents of other states, it was assumed by many local intellectuals to be phrased under the influence of Manipur-based Hmar National Congress (HNC), formed in 1950, and Hmar National Union (HNU), formed in 1958.

With the passage of time, the HPC leadership had come to realise that 'it would not be possible for them to curve out some parts of Assam, Manipur, Mizoram and Tripura to form Hmar District' (Paul, 1996: 3). So, they felt it imperative to immediately change and restrict their demand area to the Hmar majority area of Mizoram. In order to press their demand for autonomy, the HPC had, since April 1987, engaged itself in an armed conflict to pursue their contested separate ethnic identity. The rank and file of the Hmar Volunteers Cell (HVC), an armed wing of HPC, had abducted tea executives and triggered off a spate of extortions in the Hmar-inhabited areas of the three states, namely Mizoram, Assam and Manipur. The HPC's movement had turned more violent when confrontation took place between the HPC volunteers and MAP forces at Moniarkhal of Cachar District on 16 May 1989.

Although the HPC was initially formed with a view to promoting the socio-economic prosperity of the Hmars as an ethnic group, its identity politics had unleashed ethnic conflicts having far-reaching adverse effects on the development of Mizoram state in general and their inhabited area in particular. In other words, ethnic conflicts promoted by identity politics had serious impacts on development indicators in the newly formed Mizoram state.

(a) **Economy and domestic product of the state**: Since the kidnapping of tea executives by the HPC, *Bandhs*, blockage of the National Highway and ambushing of the patrolling Mizoram Armed Police (MAP) were the regular features of the days. The HPC had also blasted Mizoram State Transport bus near Vairengte village on 28th September 1991 in which four civilian passengers were killed (Lalthanpari, 2014: 367). According to the official record of Mizoram Government till July 1992, the total number of persons killed and injured in the various encounters of the HPC/HVC and MAP stood at forty-six and sixty-six, respectively. Among those killed were seven policemen, twenty-two HPC/HVC militants and seventeen civilians (Superintendent of Police, Aizawl District, 1992). All these incidents had frightened the people living in the northeastern side of Mizoram to constantly engage themselves in their agricultural activities, thereby affecting their agricultural economy. As long as the agricultural economy of the state was affected, the growth rate of Mizoram state's domestic products was surely affected.

In brief, after nine rounds of peace dialogue under ceasefire officially beginning from 31 July 1990, the HPC representatives and the Government of Mizoram had signed a Memorandum of Settlement (MoS) at Aizawl, the capital headquarter of Mizoram, on 27 July 1994, which officially ended the 8 long years of HPC movement. The Accord had provided for the setting up of '*Sinlung* Hills Development Council (SHDC) in an area 'to be specified within the HPC Demand Area of Mizoram' (Memorandum of Settlement, 1994: 4). Subsequently, 308 HPC militants surrendered along with their arms. The signing of the Peace Accord was followed by a series of meetings of the HPC representatives and the Government of Mizoram to discuss the process of implementation of the MoS. However, those meetings could not reach any agreement on the areas to be covered by the proposed SHDC for long. Ultimately, it was agreed that the SHDC would comprise the areas where Hmars made up a majority of the population. Accordingly, SHDC was officially formed on 27 August 1997.

(b) **Power and electricity as well as industrialisation:** Unfortunately, insurgent activities continued unabated even after the signing of MoS and the subsequent formation of SHDC in 1997. The reason is that a section of HPC cadres who were not satisfied with the process of implementation of the MoS had broken away from the original HPC and formed HPC-Democratic way back in 1995 to spearhead a new wave of political movement for self-government in the north and north-eastern part of Mizoram.

While the energy requirement of the people in Mizoram state was increasing day by day, the HPC (D) militants had kidnapped six employees of the Shillong-based North Eastern Electric Power Corporation Ltd (NEEPCO) from the Dam site of 60-MW Tuirial Hydel Project on 31 March 2001 (*Oriental Times*, Vol. 3 Issues 47–48, 2001) and demanded ransom of one crore for their release. Fortunately, the six captives – S. Dhar, Robert Lalsanga, Ratan Roy, Nipen Nath, Lalrinkima and D. Mandal – were released unharmed after being in captivity for 74 days. It is to be noted that NEEPCO has been undertaking the construction works of Tuirial Hydel Project and 210-MW Tuivai Hydel Project while Power Grid Corporation, a Government of India undertaking, specialised in power transmission lines, is also looking after the transmission works (ibid.: 1).

As long as generation of power and electricity was derailed by ethnic conflict, there was practically no scope for promotion of industries in and around the HPC's demand area.

(c) **Human resource development:** Though identity politics was spearheaded by HPC for the development of Hmar ethnic group, Government servants – Central and State – posted in the HPC's Demand areas had felt insecure under the horror of ethnic conflicts and left the Demand areas on ground of their physical safety. Though accurate empirical data are not readily available to substantiate the ground reality, the fact remains that some of the teachers – Hmars or non-Hmars – working in the Demand area from other places had to leave their schools for safety. Eventually, HPC (D) insurgent activities had derailed not only human resource development programmes but also other development works, including construction of communication networks in the north and northeastern parts of Mizoram.

From the beginning of 2007, the Government of Mizoram, with understanding of the gravity of the situation, had initiated a series of steps to bring the HPC (D) to the negotiating table. On 11 November 2010, after reciprocating their views, the representatives of HPC (D) and the Government of Mizoram had signed a Suspension of Operation (SoO) Agreement at the State Guest House, Aizawl, and proposed to hold the next round of talk sometime in the month of January 2011. However, no peace dialogue could take place as slated due to the question raised by the Government of Mizoram against the inclusion of non-Mizoram citizens as representatives of the HPC (D). Though the last peace talk was held sometime in the month of November 2013, no amicable peace settlement could be made by the representatives of both the HPC (D) and the Government of Mizoram.

Conclusion

Mizoram is one of the states of the region which has been victimised by a variety of conflicts. However, Mizo-Riang ethnic conflicts erupting since the post-Mizoram Peace Accord has not been dealt with in this piece of work because of the fact that Riangs are not accepted by the Mizos as the indigenous people of Mizoram. In fact, they are recognised by the Mizos as the original inhabitants of Tripura. Hmars, on the other hands, are recognised as one of the largest tribes constituting the term 'Mizo'. It is a popular belief among the Mizos that 'had the Mizo Union stood fast on its demand and succeeded in their demand for creation of an enlarged Mizo District which would integrate the Hmar majority areas in southern Manipur, Cachar and North Cachar Hills of Assam under it, the Hmars might be able to bury their ethnic nomenclature and accept "Mizo" as their identity'. However, the current ethnic divide between the Hmars, fighting for *Hmar ram* (Hmar land) under the banner of Hmar People's Convention (HPC), and the larger Mizo society composed of other Mizo tribes, including those Hmars living in the different parts of Mizoram from the time of their forefathers, appears to be initially implanted with some ulterior motives by a handful of Hmar elites within Mizoram, if not from outside Mizoram.

Most of the Mizos have realised from their experiences during MNF insurgency that conflicts could halt many development programmes. When the HPC movement was also reaching its climax, development works could not make any headway in the Hmar populated area in Mizoram. Though identity politics was initiated by the Hmars for pursuing their own development, the process of infrastructural building carried out by the Union and the State Governments in the Hmar-populated area suffered a serious setback. It will not be an exaggeration to point out that most of the development indicators which have been discussed earlier were, at a particular point of time, adversely affected by ethnic conflicts unleashed by identity politics of the Hmars who asserted themselves as a separate ethnic group. Alongside of Hmar ethnic politics, there are few other ethnic groups promoting ethnic consciousness among themselves in Mizoram. However, these ethnic groups remain dormant and do not articulate any concrete political demand for the fragmentation of Mizoram state.

While the Union Government is aiming at building a strong nation through integration of diverse ethnic groups within India, some of the ethnic groups in Mizoram are searching for an opportunity to assert their ethnic identities for petty political gains. If the Central Government is committed to national integration, it should first take concrete steps to integrate, through the state authorities, smaller cognate ethnic groups to their main tribe. In the case of Hmar ethnic politics, the State Government should listen to their grievances and accordingly take positive steps to solve their problems at least by empowering 'Sinlung Hills Development Council' to merit the status of Zilla Parishads functioning in other parts of the country.

References

Assam Secretariat (1890). *Pol. & Judicial (Judl), A., For, Progs., August 1890*, No. 1–46.

Barkataki, S. (1964). *Tribes of Assam*, New Delhi: National Book Trust.

Centre for North East Studies and Policy Research (C-NES), New Delhi, *NER Vision 2020 for North Eastern Council: Participative Planning on Development Needs (Mizoram* Chapter).

Chatterji, N. (1975). *The Mizo Chief and His Administration*, Aizawl: Tribal Research Institute.

Chaube, S.K. (1973). *Hill Politics in North-East India*, Calcutta: Orient Longman Ltd.

Cheung, Y.W. (1993). 'Approaches to Ethnicity: Clearing Roadblocks in the Study of Ethnicity and Substance Abuse', *International Journal of Addictions, 28*(12).

Das, Samir Kumar (2007). *Conflict and Peace in India's Northeast: The Role of Civil Society*, Washington: East-West Center.

First Memorandum of HPC addressed to Shri Rajiv Gandhi, Prime Minister of India dated 21 January 1987, with a subject, 'North Eastern Area (States) Re-organization'.

Government of Mizoram (1990). Report on socio-economic review 1979–90, Directorate of Economics & Statistics, Aizawl: Government of Mizoram.

Government of Mizoram (2010). 'Hydro Electric Power Policy of Mizoram', Power & Electricity Department.

Government of Mizoram, 'Mizoram: Some Facts', Directorate of Information, Public Relations and Tourism, Aizawl (pamphlet with no year of publication).

Government of Mizoram, 'Economic Survey, 2007–08', Planning and Programme Implementation Department, Aizawl.

Government of Mizoram, 'Report on Socio-Economic Review 1979–90', Directorate of Economics & Statistics, Aizawl.

Government of Mizoram, *Statistical Handbook, Mizoram, 2000*, Aizawl: Directorate of Economics & Statistics.

Government of Mizoram. *Statistical Handbook-Mizoram, 2010*, Mizoram, Aizawl: Directorate of Economics and Statistics.

Haokip, George T. (2012). 'On Ethnicity and Development Imperative: A Case Study of North-East India', in *Asian Ethnicity*, Vol. 13, Routledge, Taylor & Francis Group.

Hassan, M., Sajjad (2006). 'Explaining Manipur's Breakdown and Mizoram's Peace: The State and the Identities in North East India', in Crisis States Programme Working Papers, Series No. 1, Working Paper No. 79, February.

Karna, M.N. (1998). 'Socio-Economic Aspects of Ethnic Identity in North-East India', in R.K. Purkayastha and Gurdas Das (ed.), *Liberalization and India's North East*, Commonwealth Publishers Pvt. Ltd.

Lalthangliana, B. (1975). *History of the Mizo in Burma*, Mandalay: History Department, Arts and Science University.

Lalthanpari (2014). 'Hmar Movement for Autonomy', in Jangkhongam Doungel (ed.), *Autonomy Movement and Sixth Schedule in North East India*, Guwahati: Spectrum Publications.

Lewin, T.H. (1869). *Hill Tracts of the Chittagong and the Dwellers Therein*, Calcutta: Bengal Secretariat Press.
Lewis, A. Coser (1956). *The Functions of Social Conflict*, New York: Free Press.
Liangkhaia, Rev. (1976). *Mizo Chanchin*, Aizawl: Nazareth Press.
Mackenzie, A. (1884). *History of the Relation of the Government with the Hill Tribes of the North-East Frontier of Bengal*, Calcutta: Bengal Secretariat Press.
McCall, A.G. (1949). *Lushai Chrysalis*, London: Luzac & Co.
Memorandum of Settlement between the Government of Mizoram and the Hmar People's Convention (HPC), Aizawl, 27 July 1994.
Misra, S.K. and Puri, V.K. (2007). *Indian Economy*, Mumbai: Himalaya Publishing House.
MNF Memorandum submitted to the Prime Minister of India on the 30 October 1965.
M.U. Memorandum dated 3614/1947, submitted to His Majesty's Government, Government of India and its Constituent Assembly.
Neitham, Lalremlien (2011). 'Hmars Struggle for Autonomy in Mizoram', www.ritimo.org (accessed on 21 January 2014).
North East Vigil (2000). Issue No. 1.22, Insurgency, April 16.
Oriental Times (2001). National Vol. 3, Issues 47–48, 22 April–6 May (front page).
Pachuau, Rintluanga (1993). 'Calculation from Topographical Maps No. 84C/13, No. 84D/14, No. 84/A6 and No. 84E/6', *Vide Mizoram Science Journal.*
Paul, N. Chonjik (1996). 'Hmar Political Awakening in Mizoram', MPhil Dissertation submitted to the Department of History, NEHU, Shillong.
Phinney, J. (2003). 'Ethnic Identity and Acculturation', In K. Chun, P.B. Organista, and G. Marin (eds), *Acculturation: Advances in Theory, Measurement, and Applied Research*, Washington, DC: American Psychological Association.
Rajagopalan, Swarna (2008). *Peace Accords in Northeast India: Journey over Milestones*, Washington: East-West Center.
Records of Directorate of Economics and Statistics, Government of Mizoram.
Records of Public Works Department, Government of Mizoram.
Shakespear, J. (1912). *The Lushei – Kuki Clans*, Part-1, London: Macmillan & Co., Ltd.
Sharma, M.L. (1985). 'Ethnicity and Regionalism in North-East India – Problems of Multiple Identities and Inter-Elite Conflicts', In Pant, A.D. and Gupta, S.K. (eds), *Multi-Ethnicity and National Integration*, Allahabad: Vohra.
Simpson, J.A., and Weiner, E.S. (1989). *The Oxford English Dictionary* (2nd ed., Vol. VII), Oxford: Clarendon Press.
Singha, Komol (2012). 'Identity, Contestation and Development in North East India: A Study of Manipur, Mizoram and Nagaland', *Journal of Community Positive Practices*, 3.

Soppit, C.A. (1898). *A Short Account of the Kuki-Lushai Tribes on the North-East Frontier Districts*, Shillong: Assam Secretariat Press.

Superintendent of Police, Aizawl District, *Letter No. CRM (A)/55/92/6445*, Dated 31st July, '92, addressed to Addl. District Magistrate (J), Aizawl District, Aizawl regarding "List of HPC Surrenderees/Killed/in custody as on 28.7.92".

The Telegraph (2012). Calcutta, India, Wednesday, 22 February.

Thanga, L.B. (1978). *The Mizos: A Study in Racial Personality*, Guwahati: United Publishers.

Todaro, Michel, P. (1977). *Economics for Developing World*, London: Longmans.

Wilson, Flona and Bodil, Folke Frederickson (1991). *Ethnicity, Gender and the Subversion of Nationalism*, London: Frankcass & Co Ltd. in association with The European Association of Development, Research and Training Institute, Geneva, p. 2, as quoted in Pranami Garg's Paper on 'Aspiration for an Ethnic Identity – Evolved or Created' presented at the Seminar on '*Ethnic Identity, Social Formation and Nation Building with special reference to North East India*, Department of Anthropology, Dibrugarh University, Dibrugarh, 26–27 March 2007.

Woodford, Kate and Jackson, Guy (eds) (2003). *Cambridge Advanced Learner's Dictionary*, Cambridge University Press.

Zawla, K. (1976). *Mizo Pipute leh An Thlahte Chanchin*, Aizawl: Hmar Arsi Press.

Chapter 9

Identity, conflict and development

A study of the Borok community in Tripura

Mohan Debbarma

Introduction

An attempt has been made in this study to show that the concept of identity is a cultural, linguistic and historical aspect. Further, the interconnectedness among the concepts of identity, conflict and development has also been discussed in this chapter. With the growth of cross-border Bangladeshi migrants, the Borok community – erstwhile a dominant community in Tripura has become a minority and has been pushed to the fringes of society in political, social and economic spheres in their own land. Ethnic identity issue and the political instability of Boroks are creating a challenge in the continued efforts to overall development of the community. Keeping this in mind, the Borok community has been taken for the study to explore how the ethnic identity conflict ransomed overall development of the community in the last few decades. The study is both descriptive and analytical in nature.

Tripura is one of the small land-locked states in the Northeast region of India, covering an area of 10,491 sq km. It is bounded on the northwest, south and southeast by Bangladesh and the Indian states of Mizoram and Assam on the northeast. About 60 per cent of land is a vast expanse of lush green rain forests and the remaining 40 per cent of the land is plains. The topography does not offer any natural barrier to migration and movement and consequently becomes an important factor for the ethnic identity conflict and development stagnation owing to migration of foreign nationals.

It may be appropriate to mention that Boroks are the aboriginal community of Tripura (Debbarma, 2014: 25). They are also known as Tipra, Twiprasa, Tipperah and Tripuri. Suniti Kumar Chatterjee – a renowned historian – referred to them as the Tipperah or 'Southern Bodos' (Chatterjee, 1974: 46). Borok community comprises nine different subtribes namely Bru (Reang), Debbarma, Jamatia, Koloy, Murasing, Noatia, Rupini, Tripura and Uchoi. They share about 80 per cent of the total indigenous tribal population of Tripura. Kokborok is their mother tongue, which belongs to the Tibeto-Burman language family. In a narrow sense of the term, the Kokborok-speaking people call themselves as Borok; they consider all other indigenous tribal people such as Hrangkhawl, Molsom, Darlong, Halam, Bongcher, Kuki, Kaipeng, Ranglong, Chorai, Mog, Garo, Chakma, Khasi, Lushai, Bhutias, Bhil, Munda, Orang, Lepcha, Santal, Chaimal and the like as 'Hani Bwsa' meaning indigenous people. Boroks are the first settlers of Tripura (Debbarma, 2008: 26).

It is a historical fact that the partition of the undivided India into India and Pakistan and also the final partition of Bengal into East Bengal and West Bengal in 1947 opened the floodgates of an influx of the foreign nationals particularly the Hindu Bengalis from East Bengal (East Pakistan or present Bangladesh). This has changed the demographic scenario of the erstwhile princely Twipra (Tripura) kingdom. The genesis of the ethnic conflict between the indigenous tribal people, particularly the Borok, and the non-tribal people, particularly the Bengalis, is the continuous inflow of the illegal Bangladeshi migrants, their settlements in Tripura through dereservation of Tribal Reserved Land and socio-cultural and linguistic hegemony on the indigenous tribal people. This inflow of the illegal migrants massively continued till 24th March 1971. Within a period of 24 years, that is from 15th August 1947 to 24th March 1971, the number of foreign nationals swelled to 609,998 officially, besides lakhs of unrecorded migrants (Bhattacharya, 1988: 7). No protective measure was taken both by the State Government and the Union Government to prevent the process of unprecedented influx of illegal migrants into the state. Despite many official measures, till today, the process of influx of illegal migrants is continuing and this is one of the major problems and causes of ethnic conflict. Due to this state of affairs, the indigenous tribal community especially the Borok community has been reduced to a minority community by the non-Borok people in their own ancestral homeland since 1951. As per 2011 population census, their population share was hardly 31.80 per cent of the state total. Now, the Borok people – erstwhile a dominant community and having ruled their own kingdom for ages have been marginalised by the Bangladeshi illegal migrants. Because of this dominant nature of the Bangladeshi migrants, the indigenous tribal people have been neglected and suppressed in all walks of life, which ultimately resulted in identity contestation of the space and resources between the two ethnic groups.

In reality, the Borok population has declined from 50.09 per cent in 1941 to 38.55 per cent in 1951, and further reduced to 31.80 per cent in 2011. On the other hand, the share of Bengali migrants has increased multiple fold and become a dominant group. This demographic dynamic has paved the way for a fierce ethnic conflict that has ravaged Tripura for about last five decades. This ethnic conflict also has led to the ransom of development in the state. In addition to this, after the Indian Independence, Tripura lost its direct geographical link to the Indian mainland and this isolation stunted possibilities of its rapid economic growth despite the availability of key resources like natural gas and forest resources.

It is generally said that Tripura is one of the few places in the world and the only state in India in the twentieth century whose aboriginals have been transformed from being a numerical majority and ruling community into a minority with almost no economic and political influence in their own ancestral land. This state of affairs has threatened the very existence of the Borok identity in Tripura. As a result of which and also due to distortion of the Tripura history, the Borok community has adulterated their identity to the Bengali-dominant community. Besides, because of this foreign dominion over the natives and also due to racial discrimination and unemployment problems, some indigenous youth have taken up arms for freedom from exploitation and suppression, demanding their birth rights and rights of self-determination. Today, this seems to be the crux of the ethnic identity conflict in Tripura, which has slowed down the normal development process in the state.

Theoretical framework

The role of identity in conflict zone is adequately highlighted in the Enemy Systems Theory (Vamik, 1991: 31) and also in the analysis of Edward (1991). A synthesis of these two views brings out a persuasive concept that a divergence of basic identity manifests itself in an 'us and them' syndrome, wherein identity and negative identity play an important role in the conflict. When a group is denied land right, physical and economic security, cultural, linguistic and historical identity, political participation and recognition, leading to loss of identity in their own ancestral homeland, such a group will do whatever possible to regain it. In Tripura, there was, and still is, a divergence in the basic identity between the indigenous tribal way of life and the non-tribal way of life. Due to this divergence of identity 'us and them' syndrome started manifesting itself with the settlement of a large number of Bengalis and other non-tribal people in the state. Meanwhile, a negative identity emerges when the Boroks lost control over their ancestral land and forests to non-tribals coupled with the insensitivity of non-tribals to the local language and culture. This set the tone for ethnic identity conflict negatively affecting the development process in Tripura.

It is to be stated that the general idea of the Enemy System Theory according to Volkan *et al.* (1990) is the hypothesis that humans have a tendency to discriminate, which leads to the establishment of enemies and allies. The following ideas make up the Enemy System Theory. The first idea is that of identity with its associated concept of the negative identity and distinguishes an enemy from an ally. The next concept is that of ethnonationalism under which Montville (Volkan, 1991: p. 170) defines the concept of 'ethnic victimization as the state of ethnic insecurity caused by violence and aggression'. Depending on the circumstances, feuding parties often have a feeling of insecurity in their survival hence the tendency to protect it. Another part of this concept is the known premise among ethnonational groups that passivity ensures the continuation of victimisation (Volkan, 1991: p. 170). Horowitz (1985) and the Enemy System Theory, in combination, posit that the ethnicity spills over into conflict, as evident from the fact that conflict ensues when the security of an ethnic group is continuously shattered by violence and aggression leading to a fear of being annihilated; a native minority group cannot trust the guarantee offered by a majority that it will not abuse power; indigenous tribal groups or the minority groups are forced to abide by the entry of ethnic strangers for economic reasons; they later tend to regard them as strangers. When such conflict occurs, a group aims at the exclusion of parallel ethnic groups from the share of power. For instance, in Tripura when ethnic riots broke out in 1980 and shattered the tribals' sense of security, armed groups of both the ethnic groups took advantage of the heightened tensions and started an insurgency, projecting the non-tribals as 'foreigners' and asserting their own identity as 'sons of the soil' and on the other hand the non-tribals started an insurgency projecting the natives as 'tribal' or 'barbarous' and asserting their own identity as *Amara Bangali* (We are Bengalis) and also asserting Tripura as 'Bangalisthan'(i.e. land of the Bengalis). Horowitz distinguishes between ranked and unranked systems. Ranked systems are societies in which one ethnic group is in complete domination of another. Unranked systems are composed of two ethnic groups with their own internal stratification of elites and masses.

It is to be stated here that territory also plays an important role in the conflict as pointed out by the Territoriality Theory of Vasquez (1995).

It is his view that the territorial nature of human beings is intertwined with the sense of self and group and hence, if these needs are frustrated, conflict ensures. This thesis underlines the loss of indigenous tribal land to various development projects, subsequent forest laws and settlements of the foreign nationals in Tripura as some of the salient factors responsible for identity conflict and problem of development. Also the pivotal role of politics in conflict, particularly emphasised by Horowitz and Walker Connor (Connor, 1994), suggests that 'primordial attachments' are utilised by political leaders in ethnic conflict to achieve their ends, such as control of the state. Control of a state and exemption from the control of other alien people are the principal goals in ethnic conflicts, which are mass driven. Thus, in a severely divided society, normal administrative issues, which are routine in nature, assume a central place as political agenda. In Tripura, when the indigenous tribal people had lost their control over their ancestral land, administration, economy, history, population, culture, language and so on to the non-tribal alien people, their leaders started spearheading the tribal cause, fully aware that this was mass driven. For example, the formation of pro-tribal parties like the Gana Mukti Parishad (GMP) and Tripura Upajati Juba Samity (TUJS) brought unity and focus among the indigenous tribal people. Their success in mobilising the masses corroborates this thesis.

Another psychological aspect of conflict highlighted in the Enemy System Theory is the role of one's inability to mourn during the conflict. This theory notes that when a group is under threat and cannot let go of their losses, they suffer from the inability to mourn and tend to perpetuate conflict, as they do not want to come to terms with their loss. It is obvious from the conflict in Tripura that the native people in the state, led by pro-tribal political parties and pro-tribal insurgent groups, are yet to come to terms with the loss of land and the threat of loss of identity due to immigration of the alien non-tribal people. Even though the Tripura Tribal Areas Autonomous District Council (TTAADC) under the Seventh Schedule of the constitution has been enacted and implemented, their demand today is for even more financial powers and administrative autonomy, suggesting a refusal to accept their losses. As a result, ethnic violence is often perpetuated in the state in a cyclical manner.

In retaliation to this violent situation, a section of Bengali youth formed an insurgent organisation named 'United Bengali Liberation Front' (UBLF) and once started attacking mercilessly the innocent tribals returning homes from market places. The tribal insurgents operate their activities from the rural areas, while the UBLF operates their atrocities from the urban and semi-urban areas. So this atrocity is both to the tribals and also to the non-tribals. Interestingly, the State Government, mainly ruled by the dominant Bengalis, adopted a strategy to control the violent situation by branding that almost all the tribal areas were disturbed areas and were to be brought under the 'Disturbed Areas Act'. The Armed Forces (Special Powers) Act 1958 (AFSPA) was imposed on the tribal areas, but this Act was not applied to the non-tribal areas where the non-tribal extremists operate their violence on the natives. Another reaction of the State Government to the tribal insurgents was that many security forces were housed inside the school premises or in front of the school compound in the tribal areas for many years. This resulted in the gross violation of educational rights of the tribal people in the rural areas. In the name of tribal insurgency, many school teachers and other staff of Bengali community stopped attending their school duties in the tribal areas. In this way, schools of the tribal areas were neglected by the government. Thus, discrimination owing to this ethnic conflict has very badly affected the development process in the tribal areas.

The process of 'chosen trauma' is another aspect of conflict, reflecting an event through which, when a group is badly victimised, it becomes obsessive about the trauma and often feels a sense of entitlement. This leads to conflict. The facts regarding population inversion and loss of territory in Tripura were projected as the victimisation of tribals, making this progressive marginalisation the 'chosen trauma' in Tripura, a grievance that has, in no measure, been mitigated by the creation of the Tripura Tribal Areas Autonomous Council (TTAADC). Unfortunately, the ethnic conflict has not yet been solved by the creation of TTAADC because, it was not empowered by many of the important power, like land allotment power, direct financial power, law and security power. Thus, it has been regarded by some Borok politicians and intellectuals as 'tiger without teeth' because it was fully dependent on the State Government.

Olson (1990) outlined another set of psychological factors in the conflict dynamic, emphasising early socialisation in a violent environment, narcissist injuries such as negative identity, escalatory events like conversion experience and personal contact with terrorist groups. It was after 20 years of the major ethnic violence of 1980 that insurgent activity in Tripura once again peaked in 2000; and a majority of those who joined the insurgent groups were found to be from families affected by the ethnic violence, or from the families who lost their land to alien non-tribals as a result of deceit or deception. It may be pointed out that Terrell Northrup's (1989) escalation model emphasises that identity operates in conflict escalation in four stages. The first stage is *threat* which can be real or imaginary. The second stage is *distortion* which is the response to threat and aggression. The third stage is *rigidification* which is a process of hardening of one's attitude against the threat. The final stage is *collision* of party identities, which manifests in conflict. This pattern of escalation corresponds closely with the stages noticed in Tripura. For example, the Tripura tribal people becoming a minority due to illegal migrants was a threat to the existence of the identity of the native people, which seems to be in the first stage of the escalation model. After this stage, distortion of local culture, language, history, monuments and others seems to be the second stage of the model, rigidification of tribal people in terms of hardening of their attitude against the threat is the third stage of the escalation model of identity transformation and collision of the various political parties among the indigenous tribal people is the fourth stage of the model. All these four stages are noticed in Tripura.

Two major competing ethnic groups in Tripura

There are two major ethnic communities living in Tripura: (i) the tribal community and (ii) the non-tribal community. Tribal people are mostly the Borok people who are the indigenous tribal people. The tribal community comprises a large number of clans and communities. Borok people speak various dialects of Kokborok of Tibeto-Burman linguistic family. The tribal people belong to Indo-Mongoloid race. On the other hand, the non-tribal people are mostly the Bengalis who have become the dominant group in the state. The Bengali society consists of various castes and communities speaking various dialects of Bengali language of Indo-Aryan linguistic family (Gan-Chaudhuri, 2004: 24).

Subsequently, when the indigenous tribal people extended their settlement towards the fertile plains of western Tripura, they came in contact with the non-tribal inhabitants living in the plains bordering with Bangladesh. These non-tribal people migrated to Tripura from Bangladesh. They belong to both Hindu and Muslim religions. Further, it was not just for the love of Bengali culture and language or to be able to utilise its potential as the lingua franca between several tribes speaking different languages that the Tipra Manikya, King of Tripura, encouraged Bengali migration into the hill state. Bengalis helped to organise and maintain the structure for modern administration and the hardy peasantry of eastern Bengal reclaimed, through 'jungle-avadi leases', the undulating terrain for wet-rice settled agriculture that would boost royal revenues.

The indigenous tribal people were the nature lovers, slightly primitive, with animistic leanings. In stark contrast, the inhabitants of the plains, the Bengali migrants, had wider exposure and were ritualistic in nature. The convergence of two such diametrically opposite cultures obviously provided a perfect backdrop to conflict, since no effort was ever made to bring about proper accommodation and trust between the communities. The first recorded effort to resolve this conflict was brought about by the British political agent posted in Tripura, who pointed out that 'hill people were very simple, truthful and honest till corrupted by evil influences arising from closer intercourse with the inhabitants of the plains' (Choudhury, 1995).

Identity, conflict and development

Ethnicity, culture and identity are intimately connected with each other in terms of the identity formation of Borok community. The biological cohesiveness or racial traits are to be seen as ethnic attributes of the identity of the community, whereas non-biological traits such as language, religion, collective consciousness, self-identity, common customs, traditions and institutions and common pride in the ancestral land of origin are to be seen as the basis of their cultural distinctness. Culture, language, ancestral history, ecological setting and so on are the most fundamental determining elements for identity formation of the Borok community. Accordingly, cultural symbols are generally used as a means of identity assertion manifested through various movements in Tripura.

Mention may be made here that the Boroks are looking at their own distinct identity. Sometimes, it is identified with ethnic and cultural identities. The Boroks have been trying to assert their own identity by means of their history, tradition, language, dress, food habits and other things, but have not been able to assert their identity satisfactorily so far. The identity problem is spreading very fast owing to the dominant culture of the dominant ethnic community. Having seen this threat of extinction of the indigenous communities, identity assertion is a vital issue for Borok people of Tripura now. They assert their common identity 'Borok' or 'Tipra' through some of the common phrases such as 'Chini Borok' which means 'Our Borok' or 'our people'. 'Chini Borok' is a feeling of self-perceived identity of the Borok community. Their identity is deeply rooted in their language 'Kokborok' (Borok language), their dress 'Kanborok' (Borok dress), their food 'Chaborok' (Borok food), their music 'Rwchabmung Borok' (Borok Music), their history and other identities. In the Borok community, it is believed and said that *Dangaima* (first woman) and *Dangaipha* (first man) are Mongolian and they are Borok ancestors. Borok traditional religion is also said to be a source of identity of the Borok community. 'Huk' or shifting cultivation is another source of their Borok identity. Thus, the idea of identity being concrete is associated with sociocultural practice of the community. Borok identity may be understood as having common origin or race, common culture and language having distinct characteristics.

It is a historical fact that till Indian Independence mostly Muslim Bengali peasants migrated to Tripura from East Bengal. But in 1946, a large number of Hindu Bengalis flocked there because of communal riots in Mymensingh in Bangladesh. Most migrants since 1947 were Hindu Bengalis from East Pakistan/Bangladesh. Although the process of influx of the immigrants began in the early 1940s, it was considered to be negligible. But after the partition of India, especially in the periods between 1950 and 1971, there was a sudden spurt. This state of affairs has contributed tremendously towards the ethnic identity conflict and development.

Loss of kingdom and beginning of anarchy

It is also worth to be stated that ethnic conflict developed in Tripura basically after the merger agreement. Within a few months after the unnatural demise of King Bir Bikram Kishore Manikya Bahadur Debbarma, Tripura princely state faced a great administrative crisis and there was a threat both from the internal and external forces. The President of the Council of Regency was under pressure and had to opt to join the Indian Union. In the meanwhile, the Queen Kanchan Prava Devi on the pressure of the Government of India had to dissolve the Council of Regency and became herself as the sole Regent on 12th January 1948 and after about more than a year she headed the Kingdom on behalf of the minor Prince Kirit Bikram Manikya Bahadur Debbarma, had to sign the Tripura Merger Agreement on 9th September 1949. Thereafter, the administration of the princely state became a part of Indian Union from 15th October 1949 and later the state was administered by the Chief Commissioner, A.B. Chatterjee, as a 'C' category state. Thus, the demise of King Bir Bikram was followed by a period of political vacuum, chaos and confusion. The situation was nearer to anarchy due to lack of able leadership, Hindu – Muslim Bengali communal riots in East Bengal, influx of Bengali refugees, struggle for succession, occupation of Kamalpur by the Bengali Muslims and attempts at annexing Tripura with East Pakistan (Gan-Chaudhuri, 1985: 52). The Hindu Bengalis were willing to annex Tripura with Indian Union, but not with East Pakistan from where they were driven away and entered into Tripura.

According to the 'Sri Rajmala' or the Chronicle of Tripura, 184 Tipra kings ruled the princely Tripura State prior to its merger with India in 1949. It is a historical fact that before the merger agreement, Maharaja Bir Bikram Kishore Manikya Bahadur had earmarked 2,170 sq miles as Tribal Reserve. After his death on 17th May 1947, Maharani Kanchan Prava Devi, the Queen, took the responsibility to rule the princely state as Regent. In reality, after the demise of the king, the queen was a titular regent; the real ruler was an outsider, A.B. Chatterjee. He was the dewan, the Head of the bureaucracy of the state appointed by the Government of India. In 1948, the Regent Maharani's Dewan A.B. Chatterjee (vide order no. 325 dated 10th Aswin, 1358 Tripura Era or 1948 AD) for the first time, gave or opened 300 sq miles of tribal reserve area for refugee settlement. Later, more of these areas were opened to the Bengali refugees. Since then, the protected Tribal Reserved land, reserved for backward and underdeveloped tribals, were passed on to the refugees in both legal and illegal ways (Debbarma, 2003).

After the merger agreement, in order to provide shelter and land to the illegal migrants all these reserved land were gradually dereserved by the State Government (dominated by the non-natives), which also caused a serious negative impact upon the native people. The State Government allowed lacs of Hindu Bengali refugees into core tribal areas. It is also a historical fact that during the Maharaja Bir Bikram Kishore Manikya Bahadur, thousands of the indigenous tribal people gave moral support towards the settlement of Bengali migrants (in Tripura) who had been displaced due to the Hindu–Muslim riots in East Bengal/Pakistan on humanitarian grounds. But as a matter of fact, the displaced people did not go back to their land (Bangladesh) and they became a majority in population in Tripura owing to the unabated process of the alien migrants. Consequently, the indigenous tribal people had good reasons to feel marginalised as foreigners in their own land.

Demographic dynamics and land alienation in Tripura

It is necessary to point out here that in Tripura the ethnic conflict began after the laws were amended to recognise only individual ownership. Its objective was, perhaps, to alienate the tribal land without compensation or rehabilitation in order to rehabilitate the East Bengal refugees and immigrants at first and the construction of Dumbur dam (Bhaumik, 2003: 85). The 1947 Partition of the subcontinent into India and Pakistan was a turning point in ethnic relations in Tripura. A large number of Bengali Hindu refugees from East Pakistan came to Tripura in 1947 and immigrants followed them (Bhattacharya, 1988: 56). In fact, Tripura used to have Bengali immigrants even before 1947 but most of them were Muslim peasants who came in the agricultural season to cultivate land and went back after the cultivation season was over (Debbarma, 2009: 115–16). That was intrinsic to the British policy of supporting settlement on what they considered wasteland. Unfortunately, the system has changed with the arrival of the East Pakistani refugees at the Partition. Immigrants came later. Their flow stopped officially on 24th March 1971 when the census started counting those who were registered with the state and they were sent to the refugee camps and allotted land. The census of that year showed that the Tripura's population was 15,56,342 and the refugee's population was 609,998 (i.e. 39.19% of the total). That excluded persons who were born in the refugee camps and immigrants after their arrival in Tripura. It also excluded thousands of others who were not registered with the state. Besides, registration of the illegal migration was discontinued from 1957–58 to 1962–63. Around 180,000 persons are estimated to have come during these years or arranged their own accommodation with relatives and friends (Bhattacharya, 1988: 12–16). Because of them the tribal proportion has declined from an absolute majority to 31.1 per cent in 2001 (Registrar General and Census Commissioner, 2001).

After the refugee influx the focus of the state shifted to their rehabilitation. Most land were used for tribal sustenance. In 1948, the first ever farmers' cooperative was formed for them in the Dharmanagar (now Kanchanpur) subdivision. It became Swasti Samity Ltd. in 1950 and approximately 6,400 acres were handed over to them from around 600 Borok tribal families, mostly Reang (Bru). The Directorate of Rehabilitation was formed in 1950 for rehabilitating both registered and unregistered refugees, providing them shelter and opportunities for gainful employment. In the colonies built for them, each family was provided shelter, food rations, clothes, beds, utensils, cash, medical help, educational and some professional benefits and loans for business, land, agriculture and housing. Families that did not get land were paid Rs. 2,750 each (price of 1950) for purchasing it (Bhattacharya, 1988: 56–9). Then followed legal changes meant to facilitate land acquisition for their rehabilitation. It was a tragedy for the natives because most natives being illiterate and not familiar with the formal law did not register their land. The law did not recognise their right over the commons. So the land that was their sustenance and habitat for centuries before the colonial land laws were enacted became state property under the colonial principle of individual ownership. Most of them were not paid compensation for the land, though they had traditional rights over it according to their customary law (Bhaumik, 2003: 84). That resulted in massive tribal land alienation. Apart from the 6,440 acres of land allotted to the cooperative in 1950, the tribes had also lost not less than 26,101.2 hectares (64,470 acres) to rehabilitee colonies. Of the total, 5,499 hectares (13,572.65 acres) were private and the rest belonged to common land. No notification was issued or compensation paid for the common properties since the TLR&LRA 1960 had turned them into 'state property'. So the native people could not lay claim over their sustenance. Over and above these 70,910 acres, in 1981–82 the state acquired 3,697.03 acres, 1,164.05 acres of it for plain land Bengali immigrants and 2,532.98 for East Pakistani refugees. Thus, the total land used by the state for refugee rehabilitation was 74,607.03 acres, 17,269.68 acres of it was private and 57,337.35 belonged to common. It was taken from 22,394 tribals (Bhattacharya, 1988: 57–8). Much more land was alienated with no legal procedure through money lending or purchase. According to one study, the tribes lost 20–40 per cent of their land. Thus, the real number displaced by it is probably more than 50,000 persons (Fernandes and Bharali, 2010: 62, 87). Through these measures the state fulfilled its moral duty of rehabilitating the refugees but in doing so it ignored the interests of the indigenous tribal people. That is why the ethnic conflict erupted.

The same principle of common land being state property was used to acquire tribal land for development projects. In the early 1970s, the Dumbur dam submerged over 23,530.55 acres of tribal land and more land was used for its power house and the rest of the infrastructure. Most of it was tribal community land that was not compensated. Its project files mentioned that 2,558 individual land (or almost 13,000 persons) have been displaced but studies pointed out that around 8,000–9,000 families (40–50,000 persons) have been displaced (Bhaumik, 2003: 84). Of the 209,336.59 acres were known to have been used in Tripura for development projects, one-third of the area is known to be tribal commons (Fernandes and Bharali, 2010: 71). The Dumbur dam was the turning point because by 1970 the indigenous Borok people had lost 20–40 per cent of their land officially and more through money lending and other means. Thus, when the Dumbur dam was announced, tribals protested against it, but the state ignored them. That is when the tribal insurgency began. It is called terrorism for land (Bhaumik, 2003: 85). Despite their struggle, the tribes have not succeeded in recovering their lost land. As happened in other states, Tripura too has a law banning tribal land alienation. But because of their powerlessness the legal system goes against them and they have not been able to demand its return. This is how thousands of Borok and tribal families were evicted from their land in different parts of Tripura by the State Government in the name of development without adequate compensation. For example, from Hawaibari Sardurkarkari, Kamalpur, RC Nagar areas the natives were evicted. This is also another important cause of conflict. When the indigenous tribal people had lost control over their land and had been subjugated by the non-tribal people, they started losing various aspects of their identity.

In fact, ethnic identity conflict started right after the period when the natives became a minority in the 1950s. The end of princely rule and the introduction of Indian style ballot-box democracy also meant that the tribals would soon be marginalised as far as control over political power was concerned. The tribal people became a decisive minority, whereas the Bengalis became a dominant group in the state. As a result, the indigenous tribal people lost control over their identity, history, culture, economic and administrative power and other aspects to the dominant Bengali community. Owing to these various aspects of ethnic conflict between the two ethnic groups, the situation has gradually created some barriers to the development process in Tripura. This is the crux of the ethnic identity conflict in the state.

It is also regretted to mention that ethnic strife broke out on 6th June 1980, engulfing the entire state. Consequently, thousands of lives of both the indigenous tribal and the non-tribal people were lost in the riot, leaving thousands of people homeless. Thousands of houses were burnt. The entire state was under curfew for about one year. Over the years, the ethnic conflict between the Bengali migrants and the indigenous tribal people has intensified. The massacre at Mandwi, in which nearly 350 Bengalis were killed in one night on 8th June 1980, was followed by unprecedented ethnic riots in which more than 1,000 people, belonging to both the ethnic groups lost lives. Altogether 2,115 tribal people (age ranging from 16 to 70 years) were arrested and kept in jails during the riot of 1980. There was not enough space to keep them in the jail. So, many of them were kept in jute mills godowns at Agartala. Inhuman physical torture was meted out on the arrested tribal people. One Renu Debbarma (a government employee and Secretary of Panchayat from Jirania, Tripura) met immature death due to such inhuman torture by the jail employees in custody. Moreover, Assam Rifles and Border Security Force nabbed many non-tribal criminals involved in the riot on the spot during the riot and handed over them to state police authorities. But police released them at once. Many tribal police men were killed by their own colleagues during the riot. A section of the state police backed some non-tribal criminals in burning houses belonging to the tribals. In fact, at that time state police authorities seemed to have declared war against the indigenous tribal people.

Even after 60 years of the Indian independence, this infiltration process continues with state support. In order to rehabilitate and provide land to the migrants particularly the Hindu Bengalis coming from East Pakistan, thousands of the indigenous tribal people were evicted from their own land; and they became virtually landless. Tribal reserve lands were dereserved for the refugee rehabilitation without thinking about the consequence. Besides, their mother tongue 'Kokborok' (language of the Borok people) was neglected and not given any official recognition by the State Government. Tribal reserved posts/jobs were not maintained at all by the State Government. Thus, State Government's policies were partial, one sided and detrimental to the overall development of indigenous tribal people. Besides, a section of Bengali migrants had a trend to enjoy by disrespecting, neglecting, contempting, bantering towards the life style, language, culture and dress of the aborigines. In due course of time, it created high degree of bitterness on small issues between the two ethnic groups.

Mention may be made here that in 1979, the Foreigners Agitation was started in Assam. It spread to the other northeastern states. Tripura was one of the worst affected states as far as migration from Bangladesh was concerned. The TUJS asked for expulsion of all the Bengali Hindus and Muslims who had entered the state after 15th October 1949. The TUJS developed an armed wing called the Tripura Sena, led by Bijoy Kumar Hrangkhawl. He established links with the Mizo National Front who had camps in the Chittagong Hill Tract. He renamed the Tripura Sena as the Tripura National Volunteers (TNV). Meanwhile the Left Front pro-tribal measures created intransigence among the Bengalis. The Anand Marg, a right wing organisation in Bengal stepped in and formed a group called the *Amra Bengali* (We are Benaglis). Tripura was polarised into two ethnic camps. In Amarpur, 20 TUJS volunteers were beaten to death by an irate Bengali mob. In many places the police sided with the Bengalis against the tribals and only innocent tribals were arrested and not a single non-tribal was arrested during this riot period in 1980. Thousands of both tribal and non-tribal people were left homeless. This ethnic riot left a lasting scar on Bengali and Tribal relations.

Modern politics and conflict in Tripura

In the political parlance of Tripura today, the very terms 'ADC', 'Locals' and 'Outsiders' have almost come to denote a region characterised by ethnopolitical movements by the natives. Since the merger agreement of Tripura with Indian Union (1949), the Tipras/Tripuris have not seen a single decade of calm political atmosphere in Tripura. Instead, each decade saw new movements of political unrest, most of which turned to violent revolutions. Nature of violence is as follows. After coming from East Pakistan/Bangladesh the land-hungry Bengali migrants arbitrarily started occupying tribal reserve land across the state. As a result, the indigenous tribal people had to fight against the refugees, the land grabbers. The inward-looking self-definition of identity as an ethnonational entity now not only affects the people's relations with 'the outsiders', but also the inter-ethnic groups' relations within the region. The expectations to achieve economic and political liberation on the basis of ethnic groups have led to feuds between the people within the state. Although a common enemy is still strongly felt to be 'the outsiders', in the attempts to define one's ethnonationality, and in the struggle for 'autonomy' and liberation, the more powerful neighbouring ethnic groups came to be identified as obstacles. If the trend continues, we may expect to see more feuds and identity conflicts hampering development process among the ethnic groups in Tripura.

It may be mentioned here that the transition of Tripura from a nominal democracy to a full-fledged parliamentary democracy took a very torturous route over the period of 1949–72. This was principally due to the refugee influx which took place during this period, compounded by a lack of proper appreciation of public sentiment and of the economic difficulties faced by the landlocked state. When the Dewani system was opposed and demands for a popular elected government were made, the Government of India responded in 1951 by declaring Tripura a Part C State with an Electoral College, but without legislative powers, administered by the Chief Commissioner. When the demand for popular government intensified further, the Government of India responded by appointing a Council of Advisors to the Chief Commissioner on 1st December 1952. Coinciding with this extended period of transition was the influx of refugees, commencing in 1950. By 1958, almost 3,74,000 refugees had entered Tripura and massive rehabilitation was carried out by the Government of India. By 1960, nearly 70,000 families were rehabilitated, of whom 34,000 were settled in 75 colonies in Agartala and the rest in different places (Mohanta, 1945: 89). By 22nd October 1960, the rehabilitation work was complete. But the influx still continued till the liberation of Bangladesh in 1971, worsening the shortage of land. The important issue that has to be noted here is the lack of equity in the rehabilitation programme, between people displaced from East Pakistan, which added to the problems of Tripura (Mohanta, 1945: 89).

The Tripura Land Reform Act, 1960, also failed to reduce the inequalities in the distribution of land, or to prevent illegal transfer of land from tribals to non-tribals (Chakraborty, 2004: 148). Thus, during the struggle for Statehood, the land question, economic backwardness and the political aspirations of the tribals were systematically relegated to the background. As a consequence of the influx of a large number of refugees, a growing numbers of ethnocentric tribal parties mushroomed and the GMP, a Left Wing Tribal Party, gained popularity among the tribal political parties. The Debar Commission and Hanumanthiya Commission, which looked into the development of the Scheduled Tribes (STs) and Scheduled Castes (SCs), suggested a tribal compact area for development, which triggered the aspirations of the tribes, leading to a struggle for tribal self-rule. At the same time, the Congress Party, which repeatedly came into power in 1957, 1963 and 1967, due to its popularity among the rehabilitated people (Bengali migrants), also made a token attempt to assuage the aspirations of the tribes. They enacted the Tripura Land Reforms and Restoration Act in 1960 to restore alienated tribal land. This Act was ineffective and hence an attempt was made to enlarge its scope through an amendment in 1964. This attempt also remained a non-starter, as it was not implemented effectively, a deficiency that continues even today, as the statistics on land restoration demonstrate. The effort by the government to prevent tribal land alienation failed to prevent continuing land alienation, creating a sense of mistrust and betrayal among the tribals. The Left Wing party, which was known for its pro-tribal policies, also could do little to satisfy the demands of the tribes, further aggravating frustrations on issues of equity in their own land, provoking violence and tension against the non-tribals from (then) East Pakistan, after all the leaders were Bengalis. In this scenario, as most of the political parties failed to appreciate the demands of the tribals, and as the pro-tribal Left Party could not agree with the tribal demands, both in practice and spirit, the tribals felt an increasing need to have an alternative party which could sincerely further their cause. This led to the formation of the pro-tribal organisation, TUJS, in 1967 (Mohanta, 1945: 60). Being an exclusive tribal party, they had a pro-tribal policy and demanded

TTAADC to achieve self-rule by themselves alone, while the Left Party continued to pursue the demand for TTAADC by taking non-tribals into confidence. In 1971, to add to the heightened frustration of the tribes, the Bangladesh war led to the influx of thousands of refugees. This brought tribal anxieties on the possible loss of identity to a peak, further renewing their resolve for self-rule and recovery of land.

During the legislative assembly election of 1977, the Left parties, by virtue of their inclusive policies – covering Bengali migrants – fought the election on the issue of creating an Autonomous District Council (ADC) for the tribal people under the Sixth Schedule of the Indian Constitution, and secured an overwhelming majority, while the TUJS failed to make a proper mark in the political arena. On coming to power, the Left passed the bill for the creation of ADC (provisions as to the administration of tribal areas in Assam, Meghalaya, Tripura and Mizoram) in the Assembly (Mohanta, 1945: 67) and sent it to Parliament for their consent. The Government of India, however, gave consent for the ADC under the Fifth Schedule of the Constitution (provisions as to the administration and control of scheduled areas and scheduled tribes) on 18th January 1982, instead of the Sixth Schedule. The delay of 5 years complicated the issue of tribal self-rule and the state had to pay a very high price in the form of unprecedented ethnic riots in 1980. During this period of delay, from 1977 to 1982, in its enthusiasm to spearhead the tribal cause, the TUJS started a systematic and aggressive campaign against the non-tribals, demanding the deportation of those who migrated after the partition. Simultaneously, some frustrated Borok young tribal leaders also formed an underground militant wing by the name of Tripura National Volunteers (TNV), with an anti-establishment and anti-non-tribals agenda. The TUJS, on the one hand, started a political agitation demanding the right of tribal self-rule, while the underground TNV started a violent anti-non-tribal campaign by attacking non-tribal bazaars, policemen and forest officials. This created a massive reaction within a section of non-tribals, who formed of a parochial party, *Amara Bangali*, which systematically opposed all political moves of the TUJS, creating a very serious rift between tribals and non-tribal in the state.

Thus, from 1980 to 2001, Tripura was caught in an intractable cycle of violence, with a large number of innocent non-tribals and Leftist tribals, forest officials and Armed Forces personnel killed. This cyclical violence took place in two phases, from 1980 to 1991 and 1992 to 2008. On 17th August 1988, within three months of the formation of the new government, the TNV surrendered and signed an accord, which stipulated the rehabilitation of undergrounds activists, effective measures to prevent infiltration, the reservation of more seats for tribals in the assembly, restoration of land to tribals, redrawing of the ADC boundary, an economic package for tribal development and an increase in the number of ST seats in the Legislative Assembly from 17 to 20 (Paul, 2009: 148). Eventually, a hard assessment of implementation of the MoUs with both the surrendered groups would demonstrate that many clauses were not implemented properly. The deportation of non-tribals, which was an impractical expectation, and the redrawing of the ADC areas to include the tribal compact areas left out in the previous demarcation were particular cases in point. This, in turn, gave rise to further frustration, and the revival of militancy.

After the Left Front came back to power in the general election in April 1993, however, the entire All Tripura Tiger Force (ATTF), a pro-Left Front tribal insurgent group, surrendered in August, after signing an MoU with the government. Their demands included the deportation of non-tribals who entered Tripura after 25th March 1971, without valid documents; land reform; inclusion of tribal majority areas in the ADC; village police forces; an increase in the number of State Assembly seats for tribal from 17 to 26; recognition and promotion of Kokborok as their mother tongue. The ATTF and the Nationalist Liberation Front of Twipra (NLFT) focused their activities along the periphery of the ADC areas; terrorised the non-tribals and managed to polarise the demography, driving out a substantial number of non-tribals from the ADC areas. This continued till the 2000 ADC elections, when the Indigenous Nationalist Party of Tripura (INPT), a TUJS cloned, came to power in the ADC areas of the state. The reason for the change was the debacle of the Left Front in the TTAADC elections, even though they were in power in the Legislative Assembly. Due to political and ideological clashes between the tribal leftists and tribal rightists there had been a depletion of strength of the Front's hard-core tribal wing, GMP due to systematic killings of Leftist tribal leaders first by a section of the TNV and then by a section of the NLFT. Tribal votes were polarised due to the terror campaign of both extremist outfits in the interior tribal and peripheral non-tribal areas of the TTAADC. This turnaround of tribal sentiments was a watershed in the partial resolution of conflict in Tripura. Of late, the ruling political leadership (Left Front leadership) had identified the core issue, and then carried out a major revamp of the party, especially in the interior tribal areas. They gave full support to the state executive for a silent development of the infrastructure in such areas, concentrating on the road network, regrouping of tribal villages, providing government job to the dependents of those killed by the insurgents, reviving schools and renaming a large number of tribal villages. The government increased tribal welfare programmes and provided exclusive security for development work, revived the school infrastructure in the tribal interior and established security camps to ensure against a continuation of ethnic violence in the state.

Conclusion

In totality, there are various types of conflict confronted by the tribal and non-tribal people in Tripura. To briefly mention, they are of the following types: (i) Land Conflict – arising from dereservation of tribal reserve land (since 1949) for the settlement of Hindu Bengali refugees, eviction of tribal people from their own land in the name of development, setting dams for hydel project, eco-park, transfer and allotment of TTAADC land for the non-tribal people, racial discrimination in land allotment and documentation for the tribal people, transfer of tribal land to the non-tribal land and so on. (ii) Language, Culture and Religion Conflict – arising from imposition of Bengali language, culture and religion on the tribal people. (iii) History Conflict – arising from distortion and demolition of local history, erection of hundreds of statues of Bengali and other non-tribal people unknown to the land and its natives. (iv) Reservation Conflict – arising from improper maintenance of ST quotas in various departments, application of 100-point roster system even in TTAADC and so on. (v) Conflict arising from pre-planned intermarriage programme between the tribal people and non-tribal people for assimilation. (vi) Conflict arising from racial discrimination. (vii) Conflict arising from outsiders 'divide-and-rule' policy among the native tribes and so on.

It may also be pointed out here that analysis of economic statistics of the state for the period from 1995 to 2008 brings out the quantum jump in budgetary expenditure, leading to substantial improvements on the Human Development Index, increase in the road network, Gross Domestic Product, tribal welfare schemes, power generation, participation of the tribals in joint forest management programmes and rubber plantations, revival of school infrastructure in the interior areas and increases in minor irrigation projects. However, a small group of hard-core leaders of both extremist groups (ATTF and NLFT) along with their weaponry and infrastructure are reported remains intact in foreign countries. Promises of the State Government are broken to the insurgent returnees. State Government's development programme seems to be still partial on the basis of race, party politics, religion and other characteristics. A residual possibility of the cyclical violence reemerging in the state does, of course, remain. The conflict is reviving between the tribal people and the non-tribal people mainly owing to the continuation of the processes of discrimination, exploitation, transfer of tribal reserved land to the non-tribal people, de-resrvation of tribal vacant posts/jobs and so on. The key to an enduring solution lies in India's geopolitics and relations with Bangladesh, the sincerity and effectiveness of providing jobs to the young unemployed tribal youth in the state and preventing operational fatigue in the Armed Forces who have been operating for a period of very long time.

The migration and settlement of non-tribals caused a population inversion in Tripura, which became the crucial factor responsible for the identity, conflict and development. This inversion, coupled with a mismatch between the tribal and non-tribal ethos, as well as a history of democratic changeover, reinforces the concept of migration as one of the factors for ethnic violence. When the Borok tribal people of Tripura felt they were being denied economic security and equitable political participation, leading to the loss of their identity, pro-tribal parties emerged, leading to the 1980 riots. Among the outcomes of this conflict was the inclusion of the tribal language as an official language, an increase in the number of Legislative Assembly seats for tribals and the creation of the TTAADC. The TTAADC under the Fifth Schedule did not secure the trust of the tribal leaderships; and the struggle for the TTAADC under the Sixth Schedule sustained the cycle of conflict in the state.

The simplicity and friendly attitude of the Boroks towards the non-Boroks or the 'outsiders', the migration and settlement of the 'outsiders' and racial and historical discriminations and some other factors caused a severe population inversion in Tripura, which became the most definite and crucial factor responsible for the identity conflict in Tripura. This inversion, coupled with a mismatch between the ethos of the 'local' and that of the 'outsider' and the later phase of democratic changeover, reinforces the concept of migration of the 'outsiders' as one of the crucial factors responsible for the ethnic conflict. Autonomy and development are the two vital issues among the people of Tripura. Cyclic ethnic conflict is inevitable, in as much as the native ethnic minority is not given the sense of equity in the social, psychological, economic, cultural, historical and political space by the dominant ethnic majority. Identity crisis and conflicts between the two ethnic groups have caused a serious obstacle to the development process of Tripura. The gap in terms of development between the native people and the other people should be reduced by the government as early as possible. Thus, in order to make Tripura free from the present ethnic identity conflict, the indigenous tribal people, particularly the Boroks or Tipras, should be given more autonomy and safeguard for the development under the provision of the Indian Constitution.

References

Bhattacharya, G. (1988). *Refugee Rehabilitation and Its Impact on Tripura's Economy*, New Delhi: Omsons Publications.

Bhattacharya, Suchintya (1991). *Genesis of Tribal Extremism in Tripura*, New Delhi: Gian Publishing House, p. 78.

Bhaumik, Subir (2003). 'Tripura's Gumti Dam Must Go', *The Ecologist Asia*, 11(1): 85–90.

Benton, William (1966). *Webster's Third New International Dictionary*, Vol. II, Chicago: H.R. Publisher.

Chakraborty, Dipanita (2004). *Land Question in Tripura*, New Delhi: Akansha Publishing House, p. 148.

Chatterjee, S.K. (1974). *Kirata-Jana Krti*, Calcutta: The Asiatic Society.

Choudhury, D.K. (1995). *Administration Report of the Political Agent*, Agartala: Government of Tripura, p. 9.

Debbarma, Aghore (2003). 'The Fundamental Reasons of Origin of Extremism in Tripura and the Steps to Solve This Problem'. A paper presented in a seminar organised by Tripura Nagarik Mancha at Agartala Press Club, on 24th August, 2003.

Debbarma, Sukhendu (2009). 'Refugee Rehabilitation and Land Alienation in Tripura'. In Walter Fernandes and Sanjay Barbora (eds), *Land, Peace and Politics: Contest Over Tribal Land in North east India*, Guwahati: North Eastern Social Research Centre & IWGIA, pp. 115–16.

Debbarma, M. (2008). *A Handbook on the Identity, History and Life of Borok People*, Agartala: Kokborok Sahitya Sabha.

——— (2014). *Ethnic and Cultural Identity with Special Reference to the Borok People of Tripura: A Philosophical Enquiry*, New Delhi: Akansha Publishing House.

Donald, Horowitz (1985). *Ethnic Groups in Conflict*, Chicago: University of California Press.

Edward E. Azar (1991). 'The Analysis and Management of Protracted Conflict', In Volkan (ed.), Lexington, MA: Lexington Books.

Fernandes, Walter and Gita Bharali (2010). *Development-Induced Displacement 1947–2000 in Meghalaya, Mizoram and Tripura: A Quantitative and Qualitative Database on Its Extent and Impact*, Guwahati: North Eastern Social Research Centre (mimeo), pp. 62, 87.

Gan-Chaudhuri, J. (1985). *A Political History of Tripura*, New Delhi: Inter-India Publications, p. 52.

Gan-Chaudhury, J. (2004). *A Constitutional History of Tripura*, Agartala: ParulPrakashani.
Mohanta, Bijan (2004). *Tripura since 1945*, Calcutta: Progressive Publication, p. 89.
Northrup,T. (1989). 'The Dynamics of Identity in Personal and Social Conflict'. In L. Krisesberg *et al.* (eds), *Intractable Conflicts and their Transformation*, New York: Syracuse University Press, pp. 68–70.
Olson, Peter A. (1990). 'The Terrorist and the Terrorised'. In Volkan (ed.), Lexington, MA: Lexington Books.
Paul, Manas (2009). *The Eyewitness*, New Delhi: Lancer, p. 39.
Vasquez *et al.*, (1995). *Beyond Confrontation: Learning Conflict Resolution in Post-cold War Era*, Michigan: University of Michigan Press.
Volkan, Vamik D. (1991). *The Psychological Roots of Sectarian Terrorism*, Lexington, MA: Lexington Books.
Walker, Connor (1994). *Ethno Nationalism: The Quest for Understanding*, Princeton: Princeton University Press.

Chapter 10

Oral narratives and identity discourse in Arunachal Pradesh

Sarit Kumar Chaudhuri

Introduction

Arunachal Pradesh, the erstwhile North East Frontier Agency (NEFA), is the homeland of 26 tribes and a number of subtribes, most of them belong to the Tibeto-Burman linguistic group.[1] Sometimes, these tribes are classified into four broader cultural groups, depending on shared threads of commonality as well as cultural specificities reflected through their overt and covert aspects of society and culture. Significantly, folklore repertoire of all these tribes contains huge number of oral narratives linked with the intricate aspects of their origin and migration. And this has provided a base in search of their roots and also promoted discourses on ethnic identities within the frontier state. Identity discourse in Arunachal has taken a different turn with the administrative reform in postcolonial phase and gradually became one of the dominant issues since the achievement of statehood. Tribal situation in contemporary Arunachal has transformed a lot when we look at its historical context from the colonial phase; and undoubtedly various processes have been working collectively to bring such transformation among the tribes.[2] Bhagabati (2004: 177) rightly pointed that, 'the frontier territory's socio-political alignments as well as resolution of internal lines of difference to forge a polity capable of subsuming the discrete tribal segment, with its essential kin-based social organization, there were very little scope or the need to expand the network of social relations involving other tribal segments in the surrounding habitat except for certain well defined trade and barter transactions. However, over the years, since independence, the impact of many externally induced factors of change has been such as to alter the traditional model of inter-group relations substantially.' In contemporary Arunachal, we can locate various trends of identity construction or identity consciousness among the numerically dominant as well as smaller tribes or even subtribes.[3]

The present chapter will be an attempt to understand the nature and forms of such ongoing heterogeneous processes of identity construction and how various oral narratives largely related to origin and migration of various tribes have provided the basis to consolidate or for further redrawing the ethnic boundaries in a frontier state of India – sharing its border with Tibet, Burma (Myanmar), Bhutan and China.

Backdrop of the frontier state

As mentioned earlier, Arunachal Pradesh is the home to around 26 tribes with almost 100 subtribes or subgroups. According to scholars, like Dutta and Ahmed (1995: 20–1), the cultural fabric of the state consists of broadly five cultural zones, from west to east: (a) The Mon cultural zone, which covers Tawang and West Kameng districts and is inhabited by the Monpas and the Sherdukpens. The Akas, Mijis and the Khowas also belong to this zone. (b) To the east of the first zone lies the Nyishi cultural zone encompassing the Bangnis of East Kameng district, the Nyishis of Lower and Upper Subansiri districts and two other unique communities: the Sulungs and Apatanis. Other groups living in the zone include the Nas, Tagins and Mikirs Hill Miris/Nyishis. (c) The Adi cultural zone extends from the eastern part of Upper Subansiri district to the western boundary of Dibang valley district, covering all the Adi groups as well as the Khambas, Membas and Mishings/Miris. (d) The Mishmi cultural zone, covering the districts of Dibang valley and Lohit, is predominated by the three subgroups of the Mishmi community and among them lived a small community – the Zakhrings/Meyors. The other communities in this zone are the Khamptis, Khamiangs, Deoris and Chakmas and Tibetans who are settled in the plains bordering Assam. (e) The Nokte–Tangsa cultural zone is spread over Tirap and Changlang districts. The major communities of this zone are the Noktes, Wangchos and the Tangsas, with the Singphos, Sonowal Kacharis, Lisus and Nepalis being the small groups inhabiting the area.

Some scholars have tried to classify them into two broader cultural zones on the basis of religion: Buddhist and non-Buddhist. Nevertheless, the very criterion does not make much sense today as Christianity is another reality today, which has gained momentum within the last two decades. This is more so with the numerically dominant tribes, such as Adi, Nyishi, Galo, Apatanis, Wanchos and Noktes. Even, within these broader cultural groups, we can trace distinctive cultural specificities which indicate the extent of cultural plurality exit in this state.

Such cultural plurality is immediately visible through the manifestation of their material culture but these are also traceable within the latent aspects of culture and undoubtedly oral narratives, such as origin migration tales or legends and many other elements of intangible heritage of this frontier people, some of which are textualised in the two valuable books of Elwin (1958; 1959). Unfortunately, very few attempts are made on this line to explore such area of plurality in the light of the on-going scholarship in folklore research.

Processes of identity construction

The concept of identity is an emotive issue which has multiple connotations. In academics, perhaps it was Fedric Barth (1969) who first mentioned that the 'identity' question is basically linked with the boundary-maintenance mechanism. Perhaps, scanning through the available debates, we can look identity as a shared sense of history which most of the Arunachalee tribes today are trying to project by contextualising various elements of their origin and migration narratives – components of their oral literature or oral history as such along with other elements of their culture. The on-going process of identity formation has some very interesting features which can throw light on the evolving transformation process itself. On the one hand, some of the major tribes are trying to consolidate their numerical strength by an attempt to incorporate some of the comparatively smaller and homogenous tribes within their larger fold and on the other hand, in some cases, there are some tribes or so-called subtribes trying to come out of the larger generic umbrella in search of their own identities. Again, a few smaller tribes are trying to unify under a new nomenclature in order to consolidate their position. These can be perceived as an on-going process of the politics of numerical strength, which has meaning for the respective tribes in terms of having their shares in the existing power structure of the state and accruing benefits of state-sponsored developmental initiatives. What is important to underscore in this context is that the various elements of folk literature (origin and migration tales, genealogical narratives, etc.) are used in the process of redefining their identities or negotiating inter-tribe relations in the emerging context. Moreover, this whole process has led another important dimension in recent times where even within the tribes, different clans are also trying in a very focused way to bring out genealogical charts to understand the origin of various clans, tribes or even inter-group linkages which needs to be looked at from the larger process of identity consciousness. Also, the base line is contained within the tales or legends which are transcended orally from one generation to another.

It may be mentioned here that in the postcolonial context, identity question of Arunachal tribes is blurred with a lot of contested and confusing nomenclatures which emerged not only out of colonial writings but also due to lack of clarity in our census records. Grewal (1997: 1–15) wrote on this issue and has shown clearly the state of confusion remained in the name and number of the tribes during 1961 and 1981 Census and also enlisted dilemmas of exoethnonym and endoethnonym of the tribes. Citing examples from Elwin's book (1959) entitled *Myths of North East Frontier of India*, he has projected the idea how various legends of the Miji and Apatani tribes relate their linkages with neighbouring tribes. Though there are some significant changes in the nomenclature of tribes some tribes still continue to have this problem when it comes to the question of constitutional affirmation of new nomenclatures. One of the most glaring examples of changing nomenclature is the instance of the Nyishi who are so far noted as Daflas in Indian constitution, a term about which community in general have strong aversion as they felt it was derogatory for the people and their history.[4] This will surely pave the way for some other tribes who also want change in their nomenclatures.

However, this chapter is more focused on some other dimensions of identity construction among the Arunachal tribes which have already been mentioned earlier. Some examples from immediate past need to be discussed here, such as, the *Galo* (earlier known as the Galong)[5] and the *Adi* divide. Earlier Galos were perceived parts and parcel of the Adi constellation – the second largest tribe of the state but within a decade, there was a movement initiated by some of the Galo intellectuals opting for a separate Galo identity, which has strained the inter-ethnic relations to such an extent that gave birth to serious conflicts which were diffused with the interventions of eminent members from both the sides.[6] In fact, this identity discourse has a backdrop which is linked with the introduction of Adi primer which surfaced huge resentment in 1959 among the Galo students who probably perceived it as an imposition.[7] Though this movement has gained wide recognition, there are a few contested voices within the Galos themselves who still prefer larger Adi identity. It may be mentioned here that now this issue of two distinct tribal identities has received greater recognition from both the sides, which was perhaps quite crucial or inevitable for their mutual growth and peace. However, it may be noted that in order to carve out their distinct identity they have tried to bank upon their distinct migration tales, ritual beliefs and obviously, some distinct cultural traits. Two important and earlier writings (Dunbar, 1913; Srivastava, 1988) have textualised such migration narratives though both of them used Gallong or Galong instead of Galo which is accepted and used by the tribe. In the words of Srivastava (1988):

> Long ago, say the Gallongs, they live in the Tibetan region and on the fringes of the Indo-Tibetan frontier. Streams of migrating families came down from time to time, from upper areas and, in the absence of the adequate geographical knowledge, they followed the easiest track. Gradually, the lower areas gave them shelter and they slowly established permanent settlements. Marauding raiders from beyond the frontier raided their settlement very often, and, as they were, at that time, not very powerful in their military prowess, in comparison with the raiders, they had to emigrate . . . the Karka Gallongs, for example, had their original settlements at Pa-Pigru, near Tada Dege, just at the Indo-Tibetan frontier. Having migrated from this place they came down via the Bori area, through Peri, Kambang, Karbak, Boje, Bole to Yomsha. Yomsha became their next permanent settlement for decades. But with increase in population, the village could not accommodate all, and consequently, from here also different migrations took place in different directions. Lombi came to Jirigi and finally settled at Lombi; Tirbin came directly to Tirbin; Gamlin came to Kadai and then to Gamlin; Esi came by the bank of the Rimi river.

Dunbar (1913–17) mentioned that the Galos and the Subansiri Daflas (Nyishis) once lived in the Yamne Valley; and the Galos migrated by the gorge of Pasighat along the foot of the hills and up the Siemen Valley. Basar (2003) also referred about the Galo migration in following lines:

> Their (Galos) migration starts from the place called Yomsi. From Yomsi they migrated to Libo Ramo and then to Bori Bokar. Then following the route of Yombo river they reached Tadar Degi. From there, they migrated to Tumbin Kayi, then keeping Yombu river to the left they moved and reached Berb-jegar. Then, they migrated to Kuri Yami and then to Ragbi Robgo and settled there for some time. From Ragbi-Ragbo, Karko's son, Kore continued to move looking for a better place to settle. He reached and settled at Tega, then Rime and then crossed the river Hua and reached at

> Der-Geko. From Der Geko, he was about to reach Basar area but took a turn and reached Wak and settled there. Then he migrated to Bene, Pobdi and reached Pauk (now Alo). From there he moved to Pakam Panya and again moved back to Pauk. From Pauk he migrated to Hisum and settled at Darka. He further migrated to Legi-Leyi (Doji), then Ngomdir-Beye and again back to Doji. From Doji he moved to Lipu and then to Pigi (Bagra) and from there to Bolek Dego (Angu). From Angu he further went to Puhi Namchi (now Pushi Doke). Then crossing Rengi Geko, he reached Latum Bidum and then migrated to Bige Bite. It is at Bigi – Bite where his five sons, such as, Rina, Riba, Rida, Rihar (Basar) and Riram were born. Kore died at Bige-Bite. Rihar (Basar) then migrated towards Tabe Rau and reached Gori. Some of his descendents continued to move towards Lipu Nanchi and then to Gune. From there moved to Paka and from there they ultimately migrated to Nyigam, the present settlement.

Besides this migration question, Galos are also differing with Adis in many other socio-religious beliefs and practices, verbal arts and even in various other material elements of culture. Padu (2006: 24) mentioned that earlier writers, who invariably entered through the Minyong (Adi) region, could not penetrate in to the Galo areas; and very little could be known about them. Dunbar (1913–17) in his writing mentioned about the differences between the Abors (Adis) and the Gallongs (Galos). During that time hardly any anthropologist could get the opportunities to visit Galo areas and in postcolonial period also most of the outside scholars concentrated on the Adis rather than the Galos.

Concerning the migration of the Adi tribe, one can trace versions in relation to different subtribes of the Adis reflected in different ethnographic notes or accounts. Bhattacharjee (1975: 39–42) narrated migration routes of the Adis in the following lines:

> They (Adis) followed one of the tributaries of Salween and Irrawaddy towards Rima. From there they pushed on and followed the course of a river which they called Nyulum Siang which is likely one of the branches of the Lohit. This river nowadays is known as Zayul. Here they stayed for some time. This is the region where the blue and green beads so highly prized by the Adis, are reported to have been found in plenty. . . After a sojourn in Zayul they started on a track along the Chindruk. The river led them to the high pass called Dasing La in Tibet and Dajing Ego in Adi. They crossed it and came upon the Po Tsangpo which they called Namgong Siang,

most probably an adaptation of the Khampa name Langong Chu. This is a large tributary of the Tsangpo which flows through the area known to the Adis as the country of the Taro.

Taros are frequently mentioned in the Adi legend as very *cunning* people with evil powers. They are still looked upon by the Adis as their Ane-Abing or blood relations, because Tani, the ancestors of the Adis and Taros, descended from Pedong Nane, the Great Mother. The Solung Abang narrates the legendary struggle between Taro and Tani for the possession of the treasures and wealth of the world. Once Tani was thrown into Taro-Siang by Taro tied within a hollowed out log but he managed to come out unhurt.

It seems a conflict between the two people – the Tani (Adi) and the Taros – for at the end the descendants of Tani had to part with the Taros and leave the country. The Taro area may be taken to the country around the Showa monastery in Northeast Tibet where the Tsangpo girdles the Namcha Barwa range. They might have taken a direct route to Pemako or followed the course of the river. They had to cross another high pass called Sila-La (13,000 ft) in Tibetan or Sila-Ego in Adi and reach the Tsangpo called by them Ane Siang. Adi legends claim it to be the biggest river in the world.

Pemako, they found more congenial to their habits for it resembled in climate, flora and fauna to their original home in the Salwin-Mekong basin. Here, they appear to have remained accordingly, for a very long time, probably about 200 years. The population must have grown rapidly; and soon shortage of land was felt. Inter-clan conflicts started and colonisers started to move southwards. Natural calamities also precipitated the exodus. Adi tradition mentions a great flood known as Pumu which covered the whole land. All efforts to drain out the water failed till the mythical mithun, Polo Sobo, dug an outlet through the surrounding hills though his horn broke at Dembi La or Dembi Ego, a pass near the upper reaches of the Nugang River. The river flows in Pemako down the Ke Pang La and the route along it leads to Kongbo in Tibet. Places in Pemako still have Adi names which are familiar to all the Adis along the border.

On the contrary to Galo – Adi divide, another unique trend is visible in relation to the Nyisis, another numerically dominant tribe. In fact, in course of consolidation of Nyishi identity, they were in favour of incorporating a few tribes, such as Sulung, Hill Miris and Tagins, all Tani group of tribes within the Nyishi fold on the basis of commonalities in cultural traits and more importantly on the basis of origin tales which illustrate their brotherhoods. For example, the question of incorporation of the Sulungs within the Nyishi fold precisely rose on the basis of a folk tale which tells that they are basically originated from two mythical brothers. However, most of the Sulung intellectuals or educated are in favour of their distinct tribal identity because of their linguistic and cultural specificities. Of course, now instead of Sulung they prefer to call themselves as 'Puroik'. But historically it is true that they remained in much close proximity to the Nyishis who were basically their masters. It was Stoner (1950) who mentioned about this tribe and then Deuri (1982) also published a brief ethnographic account of this most marginalised children of the nature, largely concentrated in the East Kameng district of Arunachal Pradesh. Recently, the students of Anthropology, Rajiv Gandhi University, worked among this tribe and revealed their contemporary socio-economic status as well as the nature of various traditional institutions. It may not be worthy to mention that officially the institution of slavery is no more in existence and most of the masters had received certain amount for releasing or liberating Sulung families but actually plight of the Sulungs (Puroiks) has not changed significantly. Even today, *Rangbang* (one Sago plant) constitutes staple food for majority of the Puroiks, which is basically considered as the famine food for most of the neighbouring tribes.[8]

Incorporation of Hill Miri and Tagin with the larger Nyishi constellation was also initiated since long. Though Tagin didn't merge and came out in 2001 in favour of maintaining their distinct identity, the case of the Hill Miris was different where majority of them accepted this new nomenclature which they believe correct as per the original tales but they also determined to continue the celebration of Bori boot festival. Recently, this festival is celebrated in Itanagar with the active association of the eminent public figures of the Nyishi community and was reported as the Boori Boot festival of the Nyishis.[9] Some historical information can be added here on the initial Hill Miri–Nyishi identity dilemma which may help to understand the nature of the contested domains and the way it was finally resolved.

There are large number of printed documents which are available and deal with the historicity of the Hill Miri and Nyishi identity issue. The memorandum which was submitted by the All Hill Miri Students' Union (AHMSU) to the government in 2006 demanding the creation of the Hill Miri dialect in All India Radio Station; and Kamla Valley district incorporated their identity question also.

> It is well known to all of us that all major tribes of Arunachal Pradesh have no written script to trace their past history of origin,

evolution, etc. Our myths and legends revealed that, our tribe unfortunately lost its written script on the way which was well documented on the skin of animal gifted by the mythical father *Aab Teen*. Therefore, the *Aab Teen* sons subsequent generation had to rely on oral literature which is mainly propagated through *Nyib* (priest) and *Nyigom* gifted with stupendous memory power from Almighty who act as the custodian of the history related to our valued age old customs and traditions . . . the history of origin and migration of Hill Miri is quite fascinating . . . As per mythology, there was a big tree of large boughs. It fell down on the earth (*Chutu*) whose root sprung on the Earth. He meets a beautiful girl and lived with her and their union has given birth to *Tuni* and *Tuki*. The former son *Tuni* was *Aeb Teen* (*Abo Tani*), the forefather of Tani group of people in modern Arunachal Pradesh. *Aeb Teen* lived at the far north of earth called *Nyim* or *Suppu Rigo*. His wife gave birth of seven sons in seven years. All the seven sons were crying for there was no food for them. So, one day, while they were sleeping, *Aeb Teen* and his wife fled away down the *Nyipak* (lower plain region of the earth). The seven sons awake one by one each desperately trying to find out the foot print of their parent. Each son followed different routes through valley of different rivers and finally settled in its respective present habitat. The fourth son called *Nyi-Nyin* from where Hill Miri were born who had traced its route through river Subansiri reached up to plain of *Nyipak* (Assam) and finally settled in present Kamla and Subansiri Valley.

Hill Miris claimed that they are the descendants of Abo Tani who was having seven children. One of them was *Niyi-Niri* from whom Hill Miris were born and therefore, they are the grandson of Abotani from whom rest of the Tani tribes are also originated. Hill Miris also believed that Boori-Boot festival is one of their identity markers and even their myth about the origin of universe is distinct from rest of the Tani tribes. Their creation myth is given below:

According to their creation myth entire cosmos was once empty. In course of time *Himb* formed in that emptiness in an undefined form. From *Himb* slowly and gradually all heavenly bodies were formed. Of those heavenly bodies *Chit* (the earth) and *Dot* (the sky) were in close contact. They were separated apart with the help of a column known as *Sichi-Nyido Gilli* (centre point of earth and sky). After earth and sky were separated, the space in between was filled

> up with air and *Eji* (sun), *Hei* (water) and *Hingo polo* (the moon) were created. Thus the configuration of universe was completed. When all the heavenly bodies, the Sun, the Moon and the Earth attained their maturity, the first living being on the earth called *Bir-Bima* was born. *Bir-Bima* gave birth to *Yirkum Tami* who again gave birth to *Opo-Ane* from whom numbers of supernatural powers were born in two phases. In first phase, *Si* (the authority of the whole world), *Bur* (the God of the river and ocean), *Peka* (protector and destroyer of mankind), *Pirte* (the god of domestic property) were born in the first phase. In the second phase the *Opo-Ane* gave birth to spirits such as *Ichi* (which causes epidemic in human), *Duili* (which causes enmity and bad feelings), *Pomlejele* (which causes natural calamities) etc. After *Bir-Bima*, next living being to come in to existence on the earth was *Chutu-Ane* who gave birth to *Abotani*.[10]

On the issue of incorporation of Hill Miri in the Scheduled Tribe list by maintaining separate identity from the Nyishis, one resident of Raga, Lower Subansiri district, had filed a writ petition in the Gauhati High Court. However, both the State Government and the private respondent presented their separate affidavit-in-opposition in this case. The State Government has contended, inter alia, that there was no Hill Miri tribe recognised as one of the Scheduled Tribes in the said list under part III as per the provisions of the Constitution (Scheduled Tribe) (Union Territories) Order of 1951. So, the petitioner's contention was misleading. Judge basically said in the verdict that the petition did not have any locus stand though the matter would remain open to be agitated by right person in right forum in an appropriate manner.[11] It may be mentioned here that *Pei* clan of this tribe formed an association in 1993 called the All Pei Welfare Association to promote unity and fraternity of Pei clan to cultivate and preserve custom, traditions and rich cultural heritage and so on.[12]

Towards a unified identity

Another interesting process of identity formation may be cited here in the context of emerging realities of the state. In West and East Kameng, the Mijis (who prefer to call themselves as Sajolang) and the Akas, though represent historically two distinct tribes (both colonial documents and postcolonial ethnography mention this distinctiveness), have marriage ties which is problematic while dealing with the notion of

'Tribe' as an endogamous group. It may be mentioned here that there are lot of commonalities that can be traced, especially in their dress and ornaments. Of course, traditionally, the Aka women could have been distinguished by their tattooing marks on the face which the Miji or Sajolang women do not have. However, both of them have small population strength and are located in poorly communicated villages. They have planned for merger under a common nomenclature which is yet to be decided. They thought that with the integration they might be in a better position to articulate their unheard voices. So in order to legitimate such claim they have to depend on origin tale which says that they had descended from the same ancestors. The tale goes like the following:

> Sajolang, the descended from the sky through a silver ladder along with the forefather of the Tibetans, named Ajanguri. They were men. While coming down, they started to quarrel for an unknown petty matter. At Busobure, somewhere located at the east of the Nafra circle, Ajanguri was badly treated by Sajolang, who was wicked at the backbone. He even did not care to onset him with sword. At which Ajanguri shocked and left for Tibet having been given advice Sajolang intellectually that he should tie his limbs, otherwise he would be notorious man in the world. Sajolang could not follow the advice properly. But after Ajanguri's departure he being repented for the quarrel, tied the legs with bead threads below the knee joins and hand joins which are called Lai jang and Dree or Techu respectively. Moreover, a knot was fastened with the hair at the forehead just to remember his last word. Then he came to the present habitation of the Mijis (Sajolangs), where he became the father of a boy and a girl. Abu Gamphe Bumo and Anai Dijang Janje were the names of the son and the daughter respectively who in later course started living as husband and wife and finally given birth to sons named Sang So, Dung So, Sangthung, Khanlong, Khayoi, Khayonjew and Bewjew. Out of these brothers, Sanso and Dong So remained at the original habitat in Nafra circle who represent the Mijis who preferred to call themselves as Sajolangs. But other brothers migrated to different places like Assam, Tibet and even Thrizino circle who became Akas.[13]

It needs to be emphasised here that both these tribes can be considered as the victim of partial isolation as well as politics of under development. Naturally, a strong sense of deprivation was embedded among them and they tried to articulate their marginal voices around 1996 by initiating a movement which gained momentum under the banner 'Bichom Movement' aiming for a creation of a new district in the name of the river Bichom which flows through the Miji (Sajolang) and Aka-inhabited area. The Joint Action Committee for the Creation of Bichom District (JACCOB) was constituted at the public referendum rally organised at Trizino circle of West Kameng district which was attended by the majority of the people from both the tribes who submitted a memorandum to the State Chief Minister in 1997. In the memorandum it was categorically mentioned that the Sajolang and the Akas are single community with common culture and traditional heritage. It added that the community ties have been affected due to the change of administrative boundaries of the districts.[14] However, this movement failed to achieve its goal and within last 10 years there are a lot of change in the inter-tribe dynamics and most importantly in the political leadership in these areas. Today, Akas are really more marginalised than the Mijis, which is clearly evident in the Aka-inhabited areas,[15] but hardly any one talks about this movement; rather there is a new initiative to create a Mon Autonomous Council which has included both the Akas and Mijis along with the Monpas, Sherdukpens, Buguns and others. But it has created another contested domain and failed to get support from other parts of the state.

Mishing-Adi identity question

Mishings are part of Tibeto-Chinese Mongoloid stock and linguistically come under the Tibeto-Burman groups located both in Assam and Arunachal Pradesh. In Assam, their population is around 587,310 and are recognised as a Scheduled Tribe whereas in Arunachal their population was about 13,591 (2001 Census). They inhabit in closer proximity with the Adis in Siang and Lower Dibang valley districts who are struggling for the same status. There are lots of contested voices regarding the identity question of the Mishings of Arunachal Pradesh which can be looked at from historical evidences. Again the very etymological origin of the word is another contested domain and there are large numbers of variegated ideas traceable since the colonial literature where the term Miri was in use to designate the same brand of people. According to Pegu (1981: 6–7), 'it may sound strange that the meaning and the origin of the word "Miri" should be open to question. Some writers suggested that the term "Mishing" derived from a combination of the word Mi- (man) and Asi (water)'.[16] Pegu further added, 'but more plausible suggestion is that the term "Mishing" derived from the name "Mishing" – an offspring of "Do-shing" who was in turn the son of Pedong-ane' (ibid.: 6–7). This expression relates to popular legends; and hence it blends a popular sentimental flavour and received more credit than it probably deserves. As legends go, the Pedong-ane or Mother Pedong gave birth to several sons, namely Do-pang, Do-mi, Do-shing, Do-bang and so on. The son of Do-pang was named Padams of today. The son of Do-mi was named Minyong and from him sprang the present Minyong group. And the son of Do-sing was called 'Mishing' who gave birth to the main stream of the Mishing people of today.[17] According to Lego (2005), the Adis and the Mishings migrated from the same ancestral places in Myan Province and Kham regions of Tibet to the Siang Valley. They have same belief, culture custom faith and so on. They have faith in Sedi (earth) and Melo (Sky) or Sedi-Melo – the creators. He added a historical account revealing that the question of Adi-Mishing identity question has become one of the important dimensions in determining politics of culture. As per his narration, though the Adis and Mishings have been living in close proximity, their relationship in late colonial period remained strained. It was so much strained that older generations used to advice the young generations not to go to Mishing areas and not to speak the truth or tell the history to the Mishings. 'But in course of time, some enlightened Adi–Mishings became conscious of the importance of the Adi–Mishing unity' (ibid.: 86–94). In 1947, a few Adi and Mishing students of Sadiya High English School formed the Adi–Mishing Students' Union with its headquarters at Pashighat. Late Daying Ering, who became the Union Minister of Government of India, was its founder president. Matin Dai, who retired as the Chief Secretary to the Government of Arunachal Pradesh, was the general secretary and Osong Ering, who retired as an Indian Administrative Service (IAS) officer, was its treasurer. The union was formed to bring the Adis and the Mishings under the same and one identity as they have same origin and culture. However, with the plebiscite at Sadiya in 1951, most of the Mishing-dominated areas, such as Jonai and Laimekuri, were annexed with Assam as Mishings were divided on the issue of joining Arunachal or Assam between the pro-NEFA and pro-Assam leaderships and in the process the Adi–Mishing Students' Union became obsolete.

The Mishings of Arunachal Pradesh are not recognised a Scheduled Tribe though their counterpart in Assam are. They have been demanding the recognition. Even in 2000, the Gaon Buras of Adi and Mishing communities of Mebo circle held a meeting and decided to incorporate

Mishing to be included as a Scheduled Tribe of Arunachal Pradesh and dropped the name of subtribe Adis rather they will be treated as the Adis only. In July 2004, the State Government decided to recommend the inclusion of Mishing tribe as Adi subtribe in the Presidential Constitutional Order, 1950 and 1951 to the Central Government. It may be mentioned here that in 2006, some of the clan members along with the president of Gasing Welfare Society, East Siang district, submitted an affidavit to the Deputy Commissioner of the same district pleading for Scheduled Tribe status but under the nomenclature Minyong tribe of the Adi on the basis of the oral history as well as genealogy which basically narrates origin of clans or brother clans.[18]

In this context, Lego (ibid.: 93) gave his opinion that Mishing of Arunachal Pradesh should abrogate the term 'Mishing' and embraced 'Adi', which Adis also should ratify. On 9 June 2007, there was a meeting of the Adi Bane Kebang (ABK) at Pasighat, where the issue of Mishing tribes of Arunachal was the main topic and following resolutions were taken where they supplemented Lego's view:

1 The committee in principle accepted the 15,000 odd Mishings of Arunachal Pradesh residing in 30 villages, spread over three districts, as part of Adi tribe.
2 The Government of Arunachal Pradesh may consider in conferring of ST status to the Mishings of Arunachal Pradesh.
3 The 15,000-odd Mishings of Arunachal Pradesh residing in 30 villages should maintain the status quo of their present territorial settlements, except during the time of natural calamity of any government instruction to resettle or rehabilitate them elsewhere.
4 Henceforth, the Adi (Mishing) of Arunachal Pradesh should adhere to Adi custom and tradition. They should practically participate in activities of Adi union/society/village/Banggo forum and Kebang.
5 The Adi (Mishing) should put all effort to check illegal immigration from outside Arunachal Pradesh in their respective areas. Any such offence shall be severely dealt with as per ABK Bye-laws. In severe cases such matter may be referred to the government for further action.
6 The ABK will maintain an annual family dossier of the Adi (Mishing); and if required they will assist the Government through ABK to empower communities/sub-communities in every districts/blocks.[19]

However, this issue is not yet fully resolved but people from both the sides are trying to negotiate the issue so that it can be resolved amicably.

Expansive Tangsa identity

Tangsas are distributed in three administrative circles of Changlang district placed in between the plains of Assam and Patkoi range with more than 100 village settlement constituted of nearly 20,000 persons. Bhagabati (2004) made a valuable study on the identity question of the Tangsas about two decades ago, which have a very significant analytical frame to understand identity question within the context of their emerging social formation. This can be mentioned in brief here.

The term has multiple connotations. Dutta (1959) wrote that 'Tangsa' means 'hill people'. However, Bhagabati (2004) expressed that 'Tang' means sacrifice and 'sa' means small which means groups which made ritual sacrifice of small things, such as pig and cattle. Bhagabati adds, 'As opposed to the Tangsa, there are certain other tribes collectively referred as the "Pangsa" who too migrated from Burma, but much later than the Tangsa groups. Besides, animal sacrifice, the Pangsa groups are reputed to have practiced even human sacrifice, a practice which they continued even in their new habitat . . . "Pang" according to a Tangsa informant, means "left behind" and "sa", stands for "small" or "the younger ones". So "Pangsa" refers to the group of tribes who were left behind in Burma and who followed the animal sacrificing Tangsas in to their new habitat at much later dates. Since they came later, they were Pangsa, in other words, "younger" or the lesser ones'. Bhagabati further adds, 'the tribe of the Pangsa group are all subsumed under the Tangsa in common parlance including administrative recognition. Outsiders seldom have occasion to know that some section of the Tangsa are also known as Pangsa . . . The Pangsa–Tangsa dichotomy is purely a matter of internal awareness of the involved tribal groups.'

Though, Dutta (1959: 1) argued, Tangsa are a large tribe comprising a number of smaller tribes, Bhagabati (2004: 180) made more analytical statement and said that the social situation in the hills was never one which offered any scope for the Tangsa to work as a corporate functional entity. The smaller units of the so-called tribes were on the other hand, meaningful social universe, each possessing most of the characteristics of the tribe, such as cultural and linguistic homogeneity, territorial contiguity, a common pattern of social organisation though not organised as a single political unit under any sort of social centralised leadership or chiefs. It is the educated youth from all the subtribes of the Tangsa who for the first time made an attempt to articulate the concept of the Tangsa as a wider identity symbol subsuming the various discrete subgroup or subtribe entities; and in the process they not only formed an association called 'Tangsa Social and Cultural Development Association' in 1977 but also started celebrating a festival called 'Mol' centrally, essentially a village-based festival. This youth organisation also started addressing various social and cultural issues which they felt detrimental to their developmental cause. They tried to assess their situation in relation to much advanced tribes, such as the Adi and Apatani, and understood the necessity of projecting a unified corporate Tangsa identity though they were not averse to leave Tutsa who were vying for a separate tribal identity (ibid.: 184–5).[20] So, Tutsas are now treating themselves as a distinct tribe of Changlang district coming out of the larger Tangsa constellation.

Towards a regional identity

In recent times, in response to the wider sociopolitical realities of a nation state and strategic importance of Arunachal in the geo-politics with neighbouring China has forged upon a regional Arunachalee identity blurring the tribe, subtribe, clan identities or going beyond the origin migration tales or legends. Bhagabati (2004: 185) wrote:

> In recent years, essentially student-led movements against the settlement of the Tibetal refugees and Chakma tribals from Bangladesh in Miao and other parts of Arunachal Pradesh reflect a strong regional and not just-tribe specific interest articulation. Likewise there are other issues, arising from time to time, when the identity-framework is in terms of being an Arunachalese. A regional, state level identity is therefore, at this moment something that can be called 'episodic identity'. It surfaces, no doubt strongly, as and when a particular episode arises, e.g., the All Arunachal Student's Union led demands for demarcation of boundary with Assam, revocation of trading and contract licenses of non-Arunachalese or expulsion of foreign nationals from the State.

In the very recent time, two other important though old issues are gradually gaining steady ground. One is the occasional claim of China over the vast tract of Arunachal including Tawang district and the other is the question of *Nagalim* or Greater Nagaland by incorporating Changlang and Tirap districts of Arunachal by the Naga insurgent groups. Again the All Arunachal Pradesh Students Union and political elites, cutting across their political identities, are coming together to oppose such demands and constantly putting pressure on the state as well as

Central Governments by organising various protest movements against the redrawing of political boundary; and these can be perceived as an expression of 'episodic identity'.

Conclusion

In the 1970s, Haimendorf (1980) noticed the kind of change the Apatanis had undergone within his frame of *From Cattle to Cash.* Today (after three and half decades of his last work) when the concepts of globalisation or liberalisation almost became the passwords to understand the wider emerging realities of a nation state, such oral narratives related to origin and migration still provide us the key to understand the on-going complex intra-tribal as well as inter-tribal identity construction processes. It may be 'expansive', 'episodic' or of other forms, within the backdrop of the emerging realities of a frontier state where state craft is actually run by the dominant tribes, though represented by numerous tribes including numerically smaller or peripheral tribes. Perhaps, this identity discourse throws some light on the complexities as well as intricate processes of tribal transformation scenario of this frontier state leaving no scope for romanticising the very concept of the tribe itself. So far the diverse tribes of this state were able to project their Arunachalee identity whenever crisis situation emerged in spite of having internal line of differentiation or contested threads to weave tribe-specific identities where origin and migration myths or legends played a crucial role in most of the cases.

Acknowledgement

I would like to acknowledge that this chapter is a revised version of a paper entitled 'Oral Narratives and Identity Construction among the Tani Tribes of Arunachal Pradesh' published by INCAA, Jhargram, in a volume *Contested Identities in a Globalized World* in 2013 (edited by A.K. Danda, Nadeem Hasnain and Dipali G. Danda). I am indebted to the editors for giving necessary permission for re-publication.

Notes

1 A few local scholars now question the validity of twenty-six tribes in the light of the constitutional recognition as well as emerging identity dimensions of the tribes from the contemporary perspectives.

2 In order to have an idea of ongoing process of tribal transformation in Arunachal Pradesh, please see the essay by Bhagabati and Chaudhuri (2008) and the book by Das (1995).

3 By subtribes we mean smaller segments or groups of tribes which have territorial and clan specificities though in the context of Arunachal Pradesh sometimes this notion of subtribes is contested giving rise to multiple connotations. Some aspects of these emerging voices are discussed in the chapter which is bearing important ramifications in relation to the identity issues.

4 Showren (2007: 9–10) refers that the tradition current among the Nyishi is that they call their northern neighbour as Nyeme while their southern neighbours as Nyepak, which also means the strange person coming from the plains – non-native or outsider. Hence, the Nyishi maintain that they inhabit in the hills between the Nyeme and the Nyepak. He added (ibid.) that they were known as Chungi or Daflas in Ahom or Assam Burunjis written by Barua (1985) and Dutta (1991). Subsequently, British used these terms to dub those unruly hills men who were dwelling in the unadministered frontiers and then onwards, colonial administrators, military officials and ethnographers started designating them various names like Duphlas, Dafla, Domphilas, Bagnis, Ni-Sing, Nisu, Western and Eastern Domphilas. During my recent field work (January–February, 2009) among the Akas, it was found that they largely refer their neighbouring Nyishis of East Kameng district as Bangnis or Dophlas. One informant from Trizino, who was once considered as one of the prominent and educated Akas, gave me a new interpretation of the term Dophla or Daphla. According to him there is a tobacco plant in their vicinity which is called 'Dopat' and the people who take or consume such plant were called Dophala. According to him they also referred Nyishis as Gidji. And according to him the meaning of the term Bangni is man which he actually expressed as *Admi* in Hindi. Later on, I had a chance to collect the photograph of the plant called 'Dopat'.

5 Though in the popular writings or writings of the Arunachal-based scholars one can only find the word Galo but in the Scheduled Tribe list they are still listed as Galong. Galo intelligentsias are trying to pursue this matter so that like the Nyisi they can also succeed for a nomenclature through which they now use in their everyday life.

6 In 2005, I was returning after completion of my field work among the Adis of both Kebang and Komsing, I had to take a seat in a shared taxi which takes passengers to Pasighat at early morning. Leaving me rest all from the neighbouring Adi villages. The very conversation inside the vehicle was symbolic reflecting internal differentiation between the tribes emerged out of specific context. Such differentiation is not unique to Arunachal rather found cutting across the ethnic or political boundaries of the states.

7 Tado (2001: 84) mentioned that the Adi dialect primer was prepared by Talom Rukbo and Oshong Ering both of whom belong to Minyong group of Adis. Initially Galos didn't react but only in 1959, some of the Galo students started protesting and NEFA administration withdrawn Adi primer. And then it led to the two distinct literary and cultural societies. Of course Adi literary and cultural society remained much more active and successfully made its presence felt in four districts. It seems that at the cognitive level a final divisive space persisted which flared

up in certain contexts. Tado (2001: 85) mentioned that the student's union election of 1990–91 in J.N. College, Pasighat, led to a clash between the Galo and the Minyong students where both the groups suffered and the incident sent shock waves throughout West and East Siang districts and soon the situation was brought under control before it spread further.

8 Government tried to rehabilitate them by making colonies but didn't succeed. Many people still try to portrait them as hunter-gathers only as if they don't know the art of agriculture which in fact they practiced for their masters. Many Puroilks accepted Christianity as their new faith and a few of them coming up receiving education. Moji Riba's documentary film titled 'Butterflies in the Mud' reveals the plight of this tribe.

9 Besides cultural programme, a mithun was sacrificed as a part of the ritual which was followed by a community feast (*Arunachal Front*, 7 February 2009, p. 1).

10 This creation myth was taken from publication by the 39th Central Boori-Boot Festival Committee, 2006, Raga. It also includes myth related to the origin of Boori-Booth festival which may be traced in many other writings on this festival or book on the Hill Miris of Arunachal Pradesh.

11 Judge mentioned, 'I believe that such question of recognition is at the discretion either to the State or the Central Governments mostly on political reasons that on any existing enforceable right as covered under the periphery of the Article 226 of the Constitution of India. It is seen that everything has remained flexible. No decision has yet been taken either to recognize or derecognize the particular community of "Hill Miri" tribe. There is nothing in the case record or in the materials before me to trace out any right in favour of the petitioner. Petitioner is not duly represented from the "Hill Miri" tribe. However, in course of arguments it was not disputed that in the upper and lower Subansiri districts there resides a community known as "Hill Miri" tribe. It is also to be noted that it is not a PIL petition, and I am sitting as a single Judge. PIL is a matter to be placed before a Division bench. It has rightly been submitted by learned counsel Mr. Son that matter of disputed facts cannot be subject matter of Article 226 of the Constitution of India' (WRIT PETITION (C) No. (AP) 2001, dated 29.01.03).

12 Pei included Murtem, Makcha, Maga, Marga, Godak (Goba, Banor), Longku, Lumdik, Golom, Sim, Gocham, Chamrak, Hui, Gab, Kicho, Kigam, Himi etc. APWA is represented by an Executive Body which is constituted by thirty members.

13 For detail of the tale one can see Dutta (1990: 52–54) where he narrated the whole tale along with a genealogical chart reflecting the common ancestry of Aka and Miji tribes.

14 In order to have a detailed idea about the movement and variegated issues raised by the committee, please see Chaudhuri (2002: 99–112).

15 In 1959, Sinha (1962) first did the field work among the Akas and brought an ethnographic account of the tribe. Recently, during January–February, 2009, I took anthropology students to work among the Akas in Palizi village. I had a chance to visit Buragaon and Jamiri

areas which were the base of Sinha's work. I also went to Trizino circle and even visited Aka village in Bana area under East Kameng district. They are locally called Koro Akas. People allege that factionalism within the tribe and crisis in political leadership are responsible for such backwardness in Aka inhabited areas.

16 Pegu cited from the writing of T. Pamegam published in *Asomer Janajati* which was published in 1962.

17 Pegu mentioned that Late Apak Jamu – Head interpreter of NEFA (now Arunachal) in British India – supported this expression. His opinion was published in 'Lolad' 1948 – a mouth piece of them 'Sadow Asom Miri Chatra Sanmilon' held at Tari gaon. There are few who think the term has derived from 'AMI plus ANSING' meaning peace loving people.

18 In brief the letter says that Tamut, Riyang clan of Minyong group of Adi from East Siang and West Siang districts are aware of the history of their clans and therefore they declare that Pao's from Oyan village, Namsing village, Gadum and Mer villages are originally our very own clan brothers related to us by blood and originally belong to Minyong group of Adis though presently known as Mishing and therefore they should be given APST status as Myniong of the Adi tribe. For detail one should look at the Affidavit given to D.C. East Siang District, Pashighat.

19 This is reflected in the official document of Adi Bane Kebang held on 9 June 2007 at Pasighat.

20 Bhagabati (2004: 185) mentioned a quotation of one senior Tangsa student which can be mentioned here, 'The Tangsas are all brothers, only our elders did not try to bring the brothers together. They were busy smoking opium and going about their own separate ways. We now want all to be together. The Tutsas want to go their separate ways. Let them. They were never Tangsa; only the Government put them with us.'

References

Barth, F. (1969). *Ethnic Groups and Boundaries*, George Allen & Unwin, London.

Barua, R.S.G.C. (1985) [reprinted and translated]. *Ahom Burunji: From the Earliest Times to the End of Ahom*, Spectrum Publications, Guwahati.

Basar, J. (2003). 'IKS of the Galos', M.A. dissertation, unpublished, Rajiv Gandhi University, Itanagar.

Bhagabati, A.C. (2004). 'The Tribe as a Social Formation: The Case of the Tangsa of Arunachal Pradesh', In A. Basu, B.K. Dasgupta, and J. Sarkar (eds), *Anthropology for North East India: A Reader*, Indian National Confederation and Academy of Anthropologist and National Museum Man, Kolkata.

Bhagabati, A.C. and S.K. Chaudhuri (2008). 'Social Transformation Process in Arunachal Pradesh', *Bulletin of the Department of Anthropology* (Special Issue: Fifty Years of Anthropology Department [1956–2006]), 10: 15–24.

Bhattacharjee, T.K. (1975). 'The Adis and their origin and Migration', *Resarun*, 1(2): 2.1–2.6.

Blackburn, S. (2003). 'Memories of Migration: Notes on Legends and Beads in Arunachal Pradesh, India', *European Bulletin of Himalayan Research*, 25(26): 15–60.

Chaudhuri, S.K. (2002). 'Marginalised in Our Own Homeland: Bichom Movement in Arunachal Pradesh', In I. Barua, S. Sengupta and D. Dutta Das (eds), *Ethnic Groups, Cultural Communities and Social Change in North East India*, Mittal Publications, New Delhi, pp. 99–112.

——— (2009). 'Inventing Images and Reforming Tribal Religion: Emerging Realities of a Frontier State', In Purvottari (ed.), *Spirit of North-East* (abstract book of a national seminar), Indira Gandhi National Centre for Arts, New Delhi.

Das, G. (1995). *Tribes of Arunachal Pradesh in Transition*, Vikas Publishing House Pvt. Ltd, New Delhi.

Dunbar, George D.S. (1913–17). *Abors and Galongs: Notes on Certain Hill Tribes of the Indo-Tibetan Border*, Asiatic Society, Calcutta.

Dutta, D.K. (1990). *'Sajolang – The Ancestor of the Mijis'*, In P.C. Dutta and D.K. Duarah (eds), *Aspects of the Culture and Customs of Arunachal Pradesh*, Directorate of Research, Government of Arunachal Pradesh, Itanagar.

Dutta, P. (1969). *The Tangsa of the Namchik and Tirap Valley*, Directorate of Research, Itanagar.

Dutta, P. and Ahmed, S.I. (1995). *People of India: Arunachal Pradesh*, Seagull Books, Calcutta.

Dutta, S.K. (1991). *Assam Burunji* (1648–1681), Govt. of Assam in the Department of Historical and Antiquarian Studies, Guwahati, p. 79.

Elwin, V. (1958). *Myths of the North East Frontier India*, Directorate of Research, Government of Arunachal Pradesh, Itanagar.

Elwin, V. (1959). *The Art of the North-East Frontier of India*, Directorate of Research, Government of Arunachal Pradesh, Itanagar.

Grewal, D.S. (1997). *Tribes of Arunachal Pradesh – Identity, Culture and Language*, South Asia Publication, Delhi.

Haimendorf, C.V.F. (1980). *A Himalayan Tribe: From Cattle to Cash*, Vikas Publishing House Pvt. Ltd, New Delhi.

Lego, N. (2000). *History of the Mishings of Arunachal Pradesh and Assam*, P. Lego, Itanagar.

Lingfa, P. (2006). 'A Brief Study on Dolo Descendants of East Kameng', Unpublished paper presented in a seminar on 'Traditional Culture and Faith of the Nyishi: Change and Continuity', organized by the Nyishi Indigenous Faith and Culture Society and Vivekananda Kendra Institute of Culture, Itanagar on 2–3 December, 2006.

Nani, Yadi (2006). 'The Migration and the Settlement Pattern of the Apatani Valley', Unpublished dissertation submitted for M. Phil Degree at R.G. University, Arunachal Institute of Tribal Studies (AITS), Itanagar.

Padu, K. (2006). 'Formation of the Organisation of Padu-Padung: An Ethnographic Study', Unpublished M.A. Anthropology dissertation submitted in AITS, Rajiv Gandhi University, Itanagar.

Pegu, N.C. (1981). *The Mishings of the Brahmaputra Valley*, M. Pegu, Assam.

Tado, P. (2001). 'Ethnic Situation in Arunachal Pradesh: An Over View', *The Journal of the Anthropological Survey of India*, 50(4): 83–8.

Tara, T.T. (2006). 'Pirya Payenam', Unpublished paper presented in a seminar on 'Traditional Culture and Faith of the Nyishi: Change and Continuity', organized by the Nyishi Indigenous Faith and Culture Society and Vivekananda Kendra Institute of Culture, Itanagar, Arunachal Pradesh on 2–3 December, 2006.

Showren, T. (2007). *The Nyishi of Arunachal Pradesh: Brief Ethnographic Outline*, All Nyishi Students Union, Itanagar.

Sinha, R. (1962). *The Akas*, Directorate of Research, Government of Arunachal Pradesh, Itanagar.

Srivastava, L.R.N. (1962). *The Gallongs*, Directorate of Research, Government of Arunachal Pradesh, Itanagar.

Stonor, C.R. (1952). The Sulung Tribe of Assam Himalaya, *Anthropos* (Sep–Dec)H.5/6:47: 947–62.

Chapter 11

Identity, conflict and development in Nagaland

Kilangla B. Jamir

Introduction

After millennia of wandering, a number of ethnic groups had settled in the hills of the Eastern Himalayan region. The Nagas are one among them, and they are concentrated in the state of Nagaland, parts of its neighbouring states and Myanmar. Major Naga tribes in Nagaland are Angami, Ao, Chakhesang, Chang, Khiamniungan, Konyak, Kuki, Lotha, Phom, Pochury, Rengma, Sangtam, Sema, Yimchunger, Zeliang and some other subtribes. Although each tribe speaks different dialects, all of them belong to the Mongoloid stock and speak Tibeto-Burman language. The Naga tribes are believed to be migrated from central part of Asia to the region in separate groups after crossing the Irrawaddy and Chindwin rivers of Myanmar, and they made their settlements in the ridges and mountain terrains of the northeastern part of India (Chatterjee, 1950).

The ancient Nagas remained in isolation for many centuries; each tribe comprised several villages with well-demarcated boundaries. Each village was a sovereign village and organised local self-government based on customary rules, usages, conventions and laws, which were mostly republican sovereignty in nature. There were inter-village feuds that often resulted in head-hunting wars among them. Their democratic spirit, courage and boldness to fight against enemies, love for freedom and liberty, social behaviour and custom-based life make their identity unique (Zhimomi, 2004).

The isolation of the Nagas was broken with the advent of British colonial administration against their wishes. The consolidation of the Naga tribes under the British rule and the spread of modern education have awakened the feeling of nationalism with growing political consciousness. Thus, the Nagas expressed the aspiration to have an independent status and to protect their identity after the departure of the British rule from India. Despite their plea, the British ruler handed over the Naga territories to India and a small portion to Burma. After the independence of India, the Naga political movement intensified leading into an armed conflict with the Indian armed force, which continues till date without a permanent solution. Attainment of statehood of Nagaland, followed by numbers of ceasefires, signing of agreements and peace talks, could not fulfil the aspiration of the people and thus, failed to bring about lasting and honourable solution to the Indo-Naga conflict. The process of economic development which was initiated since its statehood has been rather unprogressive due to numerous reasons. The objectives of this chapter are the followings:

1 To ascertain the role of British authority in awakening the feeling of the Naga nationalism and aspiration to protect their identity as an independent entity from that of India.
2 To explain how this has led to conflict with India during the post-independent period.
3 To discuss the process of economic development amidst the armed conflicts.

The information and data have been drawn from secondary sources. The growth trend in NSDP (Net State Domestic Product) has been analysed for the period 1980–2006; other decades have not been included due to the lack of data. In order to assess the impact of conflict on economic growth, this period has been divided into three sub-decades: the 1980s, 1990s and 2000s. The 1980s is considered relatively more peaceful than the 1990s. In the last decade, the situation was relatively normalised due to signing of ceasefire agreement between the Government of India (hereafter, the centre) and the underground factions. The growth trends of economic sectors in NSDP have been estimated for these periods using exponential regression method.

Role of British in shaping the Naga identity and conflicts

The first contact of the Nagas with the plain people of India was made during the thirteenth century AA by the Ahom rulers of Assam (Devi, 1968). Thereafter, some continuous relations/contacts have been made between them in the form of raids, trade for salt and clothes primarily and even matrimonial alliances. In fact, it was the British ruler who broke the isolation of the Nagas and integrated the Naga territories into the British colonial administration. The annexation of Naga territory was essentially to serve their commercial interest in the northeastern frontier, linking Assam with Manipur and Burma through a direct route via Naga territories (Mackenzie, 1969). The British government sent number of military expeditions to subjugate the Nagas from 1839 to 1850. They have fought against the British advances into its territories due to strong desire to protect their sovereign-independent status, but were defeated easily because they were not united. There were direct confrontations by the Angami Naga with the British rulers. The Nagas continued to fight back bravely against the mighty invasion, but were defeated; the battle of Khonoma marked the end of the Naga hostility against the British. The British authority brought the Naga territory under their rule by making a new administrative zone with Chumukedima as headquarter, which later shifted to Kohima in 1877 (ibid.: 125–30). A full-fledged Naga Hill District was created in 1881 (Bhatt and Bhargava, 2006); this marked a new phase in the history of the Nagas and the British as well. Later, their rule was extended to other places.

Under the British administration, the Naga Hills and its people were kept in isolation from the rest of India. The Bengal Eastern Frontier Regulations was already passed in 1873, which prohibited the plain people from freely entering the Naga Hills area. The Government of India Act, 1919, authorised the Governor of Assam to rule the hills, keeping away from legislative acts. The Government of India Act, 1935, made provisions to declare the district including other hill districts of Assam as 'Excluded Area' that kept the districts out of general administration. As such, no economic development efforts were initiated by the British government in the Naga Hills; they neither taxed the Naga people except a nominal house tax collected from each household. The most remarkable progress during this period was the coming of American missionaries who introduced modern education along with Christian religion.

The awakening of Naga national feeling emerged with the growing political consciousness of the people under the British rule. The spread of modern education and consolidations of the Naga tribes under the British rule enhanced the common feeling and sentiments among them for having an independent status, free from outside interference as before the British invasion, which became the bases of the Naga political movement in demand for political sovereignty and to protect their identity. The British officials posted in the Naga Hills paved the way to maintain harmonious coexistence among the tribes, done away head hunting practices among themselves and emerged aspiration of a separate political identity after the departure of the British from India (Anand, 1980).

The demand for independence can be dated back to 1918, under the banner of the Naga Club which submitted a memorandum to the Simon Commission in 1929 representing the aspiration of the people to be independent from India (Bhatt and Bhargava, 2006). The Naga Hills District Tribal Council was organised in 1945 with the assistance of C.R. Pawsey, the then British deputy commissioner posted at Kohima, which replaced the Naga Club, later it was renamed as the Naga National Council (NNC) in 1946. It provided a common platform for the unification of all Naga Tribes (Nag, 2013). Memorandums and petitions submitted to the British authorities expressed the aspirations of the people to regain the sovereign status after the British departure. Moreover, the Nagas had an apprehension that their identity, culture and social institutions, which are different from that of Indians, would be insecure unless they have independent political status. Further, they have correctly anticipated that they would be neglected and discriminated by the dominant Hindu and Muslim communities.

This appears to be the reality, because the Nagas as well as the entire northeastern youths who are studying and working in the mainland India are confronted with constant harassment, and they even loss their lives just because their looks and cultures are slightly different from those of mainland India. Their lives are insecure in their own country because Indians fail to accept the people from the region as one of them. The sense of 'alienation' is being felt increasingly by the common people from the northeast region. In the long run, such reality will have greater negative impact on the unity and integration of the country.

Despite the appeal of the Naga leaders for autonomy, the British authority handed over its political power to India on 15 August 1947. The Naga Hills was simply included as an integral part of Assam. Moreover, the Naga-inhabited villages and areas were broken up absurdly whereby placed a major portion of the Naga-inhabited area under the present state of Nagaland and the remaining areas were placed under Assam, Manipur, Arunachal Pradesh and Burma (Myanmar). For instance, the international boundary line between India and Burma runs through the middle of Longwa village chief's (Ang) house, so the members of the household can be both Indian and Burmese citizens.

Indo-Naga conflict in the post-independence period

On 14 August 1947, the NNC declared the Naga Hills as independent nation. That gave out a strong signal to the world that they would not give up their political right. The Naga leaders determined to fight for political sovereignty and work for the awakening of Naga political consciousness under Zapu Phizo, who later became the Chairman of the NNC in 1949. They boycotted the general elections in 1952, and in the subsequent years, the formation of the Federal Government was declared and it hoisted its flag for the first time in 1956 (Zhimomi, 2004).

Initially, the NNC adopted peaceful path by taking unilateral plebiscite, boycotts, non-cooperation and so on, but all these actions were not taken seriously by the government of India. The government thought that it was a matter of law and order situation that needs to be dealt accordingly. Unfortunately, it was not so simple and smooth. The stand of the NNC has been 'the Naga tribes has never been a part of India (neither conquered by any Indian rulers nor consent), but wholly a separate country from India historically, culturally and geographically, with an undeniable right to be recognised as politically independent' (Nibedon, 1988). On the other hand, the Indian government neither leave the Naga territories nor honour the national aspiration of the Naga people. This impasse had led to armed conflicts, which is still on for more than six decades without any permanent solution.

The centre responded by deployment of thousands of armed forces and declared the Naga Hills as 'Disturb Area'. This empowered excessive/extreme power to the army to deal with the independence movement. The unfortunate turn of the events was that those armed forces harassed the innocent villagers, tortured them to the extreme, shot at mere suspicion or without any reason, raped women, burnt houses, granary and churches to ashes and grouped or cramped a number of villages in a single village, like in war camps. Thus, the Naga political history took a dynamic turn; the general masses have supported the NNC movement; and a large proportion of volunteers were women. Political confrontation, armed struggle, uncertainty, fear psychosis, loss of lives and properties and extreme physical and mental torture under the hands of Indian armed forces have characterised the movement for independence (Zhimomi, 2004).

By 1956, thousands of Nagas joint the movement but the armed struggle appeared to be futile. This led to differences in ideology among the Nagas. Gradually, a moderate section of people, known under the Naga People's Convention (hereafter, the NPC), were in favour of autonomy under Indian union while the others continued to hold on to the sovereignty or self-determination of the Nagas. The centre responded to the demands of the NPC positively. A separate administrative unit known as the Naga Hills Tuensang Area (NHTA) was formed in 1957. This development was outrightly rejected by the NNC, and they intensified their activities against the Indian Army and NPC. The NPC further approved the Sixteen-Point proposal which became the bases for creation of the state of Nagaland, separated from Assam on 1963. Then, the process of democratic legislative assembly and the state politics started to take roots, and economic development initiatives were taken up.

During this period the armed conflict between the NNC and the Indian armed forces had intensified. Naga territory was once again declared as Disturb Area; under this situation the innocent people were once more caught in the whirlpool of fear and terror created by both the conflicting parties. The NNC by then established link with the neighbouring countries from where they received arms and training supports.

Despite being separated from Assam with the attainment of statehood, the normal lives in Nagaland were uncertain and the economic development process was jeopardised. The Baptist Church leaders urged the centre and the NNC to form a peace mission, and ceasefire agreement was signed in 1964, but it failed to bring about any solution. Subsequently, the 'Shillong Accord' was signed in 1975 at Shillong, Meghalaya, between the centre and NNC, whereby the NNC representatives agreed on their own volition to accept the Constitution of India without condition, surrender their arms and renounce their demand for the secession of Nagaland from India. It was also agreed that the representatives of the underground organisations should have reasonable time to formulate other issues for discussion for final settlement (Wikipedia, 2014).

However, many of the Naga people and the NNC leaders who were abroad at that point of time condemned the agreement that ultimately created factionalism among the rebels. The internal feuds resulted in creation of the National Socialist Council of Nagaland (NSCN) by breaking up from their old organisation NNC in 1980 under the leadership of Thuingaleng Muivah, Isak Chisi Swu and S. Khaplang. Soon, in 1988, the NSCN was further divided into two factions: NSCN (I-M) under Isak and Muivah's leadership and NSCN (K) under Khaplang (ibid.). The factional fights intensified. Moreover, in 1990, after the death of Phizo, the NNC was further divided into two factions. Further, another faction has emerged in recent years known as NSCN (U). The Naga freedom fighters are now divided into multiple factions, which weaken the movement and misled the aspiration of the people.

In the 1990s, along with the factional fights, the armed conflicts with the Indian Army still continued. Once again Nagaland was put under

Disturb Area Act. Under this act, many innocent people have undergone humiliation, and lost lives and property. On the other hand, the member of underground factions had also inflected misery to the innocent lives in the form of taxation, extortion, killings of innocent lives and others. The situation became extremely unstable and insecure for the general public. The public cry for withdrawal of the Disturb Area Act as it resulted in gross violations of human rights.

In 1997, through the initiatives of the Baptist Churches, the Atlanta Peace Meet for all the factions was arranged, but it was boycotted by the NSCN (I-M). In the same year, another ceasefire agreement was made between the centre and NSCN (I-M) and soon it extended to other factions. The civil society and the Baptist Church Council are actively trying to bring in unity among the factions. On the other hand, talks are going on between the centre and the insurgents in which Greater Nagaland by unifying all the Naga-inhabited areas seems to form one of the main agendas. However, negotiation with just a fraction of population and the final outcome, if any, may possibly lead to another dimension of conflict in Nagaland.

Development processes amidst the conflicts

In the post-independence period, the Naga territory had experienced armed conflicts, and it has always been under parallel governments, the centre on one hand and the underground factions on the other. Prior to its statehood, the administrative machinery was mainly geared to suppress the mass-based political movement. Thus, there was no significant process of economic development initiated. Moreover, Nagaland state was created under very abnormal situation. Its formation was preceded by unprecedented violence and misery of the people associated with the mass-based political movement. To assuage apprehensions of the people regarding protection and preservation of their identity, the provision of certain constitutional protection was required. As such, it resulted in Article 371(A) of Constitution of India, which gives certain autonomy to the people, and recognised their culture and its practices, yet conflict continues.

Although the cause for the conflict was not due to lack of economic progress but demand for political sovereignty, the regular conflicting situation in the past has no doubt retarded the process of economic development. For economic development and growth, political stability and conducive social environment are needed, which cannot be created where there is constant armed conflicts. Moreover, without a strong economic base, political freedom alone will not ensure true freedom to its people.

The state

Nagaland is one of the states of India located in the northeast of the country and occupies a total area of 16,579 sq km. In terms of percentage, it accounts for 0.52 per cent of the country's geographical area, bounded by Arunachal Pradesh in the north, Assam in the west, Manipur in the south and Myanmar in the east. It is almost entirely inhabited by tribal population with their individual distinct lingual and cultural features. According to 2011 census report, the total population was 19, 80,602, with a density of 119 population per sq km against the country's density of 362 per sq km. There are 11 districts and altogether, there are 1,428 villages and each village has a Village Development Board (VDB), which serves as a decision-making body as well as implementing agency for developmental works at the village level. Further, the 'Communitisation' programme is a step taken up by the government in recent years with an objective to transfer responsibilities to the people at grass-root level to manage the basic sectors such as education, electricity, water supply and health; also such initiative would inculcate a sense of belongingness in the government among the people.

Economic development

Nagaland began its first phase of economic planning in the 1960s. During the initial period the economy was based on traditional agriculture. Agriculture was the main occupation, in which 89.55 per cent of its working population were engaged (Government of Nagaland, 1973) using primitive technique of production. On close observation of development process since its statehood, Nagaland has witnessed certain changes in the structure of its economy, but has not achieved the expected level of development, rather a few people at the realm of power could amass much resources for themselves, leaving the people at the grassroots at backwater. Thus, the inequality in income and regional development within the state is becoming evident. That, in recent years, the Eastern Naga People Organization (ENPO), comprising Tuensang, Mon, Longleng and Kiphire districts, has started demanding a separate state, mainly due to slow process of economic development in that region under the current administrative set up (*The Indian Express*, 2012).

Nevertheless, development is evident in the state in respect of basic minimum facilities like education and health services in the recent past. The number of educational institutions has increased from 599 in 1961 to 2,624 of different levels in 2007–08. As a consequence, the literacy rate raised from 17.91 per cent in 1961 to 80.11 per cent in 2011. This brings in an attitudinal change among people towards progress and development. However, it lacks professional and technical educational institutions. In case of health-care services, the number of medical personals has increased from 40 doctors to 452 during 1961 and 2011, respectively, and health institutions have also increased remarkably, which led to improvement of general health conditions, and thus it reduced the death rate from 6 persons to 3 persons per thousand populations during 1980 to 2009, respectively (Government of Nagaland, 1973; 2009; 2012). The Human Development Index (HDI) rose from 0.32 in 1981 to 0.42 in 1991 and further to 0.62 in 2001. On the other hand, the Human Poverty Index (HPI) declined during the corresponding periods as 49.37, 42.07 and 35.58 (Government of Nagaland, 2004).

Improvement in road and network connectivity has also been made. The state had a total road length of 10,097 km, of which 59 per cent was surfaced in 2013–14 (Government of Nagaland, 2014). Due to difficult geographical terrain, only a small airport and the railways (only about 10.45 km) have been developed at Dimapur. According to census report, 2011, household with internet connectivity is found to be very negligible, estimated at 1.73 per cent. Also 75.22 per cent of the households have got electricity connectivity but its service is extremely irregular. Out of total energy consumption, domestic lighting constitutes about 66 per cent, commercial activities 11 per cent and industrial lighting only 3.87 per cent in 2011. This reveals that the state has got low economic activities in commercial and industrial sectors. Moreover, the living conditions as determined by housing condition show that 18.94 per cent of households still reside in thatch-roofed houses, and 52.4 per cent of household had bathroom facility. The proportion of household with tap water supply is 52 per cent and 36 per cent in rural and urban areas, respectively. But a large proportion of household depends on rain water harvesting because tap water supply is extremely unreliable.

Agriculture is still one of the major economic activities. But it is dominated by traditional method of shifting cultivation, which is at subsistence level. The efforts have been made under the government initiatives to improve this sector. As such, the irrigated area has increased from 12,587 hectares in 1961 to 83,500 hectares in 2011, but the extent of usage of modern machinery is negligible. Its total productivity was increased from 0.71 MT in 1961 to 2.61 MT per hectare in 2010–11.

Furthermore, horticulture and cash crop cultivation are initiated in the state, moving slowly towards commercial production (Government of Nagaland 1973; 2012).

The secondary sector comprises a few small and tiny cottage-scale units based on local resources. A few medium-scale industries have been set up during the initial period like sugar mill at Dimapur, paper mill at Tuli, plywood factory at Tizit and cement plant at Wazeho. Unfortunately, almost all of them have become non-functional today due to one or the other reason; otherwise, these units would have provided employment opportunities to thousands of labour force and could have contributed significantly towards economic development.

Share of economic sectors in NSDP

Although agricultural sector dominates the state's major workforce, it is the tertiary sector that has contributed around 53 per cent and 51 per cent in NSDP during 1980–81 and 2005–06, respectively. Among its sub-sectors, the transport and communication (13 per cent) has emerged as the leading sector, followed by the public administration (12 per cent) and the business services (11 per cent) in 2005–06 (Appendix A11.1).

The primary sector is the second major sector, contributing around 33 per cent and 35 per cent in 1980–81 and 2005–06, respectively. Within the primary sector, the agriculture as a sub-sector has shared 29 per cent and 31 per cent of NSDP in1980–81 and 2005–06, respectively, which is still the key sector in the state's economy as indicated in Appendix A11.1. It is not only the largest contributor to NSDP among the sub-sectors, but also the largest provider of job that engages around 61.66 per cent of total workforce according to census report, 2011.

Figure 11.1 shows that the secondary sector has remained stagnant except for 1990–91 and continued to be the lowest contributing sector in the economy, accounting for around 14 per cent during 1980–81 to till 2005–06. In recent years, the construction (11 per cent in 2005–06) has emerged as an important sub-sector, although not significant in total NSDP (Appendix A11.1).

Annual growth rates of NSDP in Nagaland during 1981–2006

The armed conflicts have been going on for several decades. But it is difficult to quantify the economic outcome of such conflicts as this requires data on the resources mobilised by the insurgent factions and

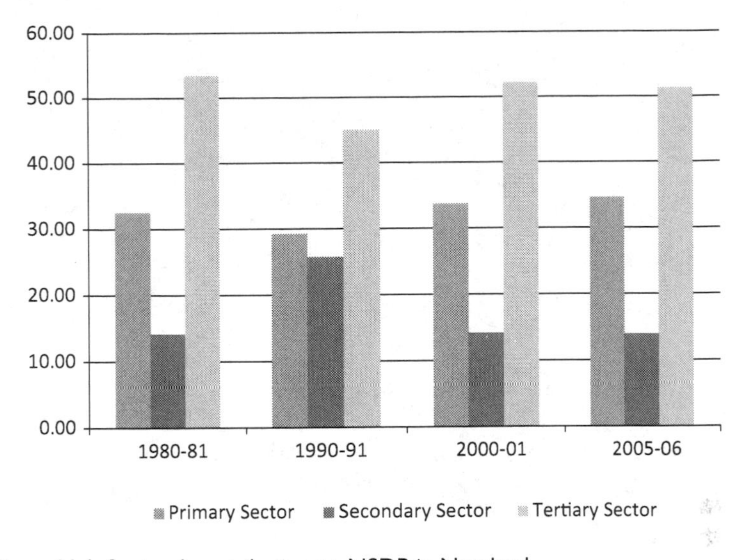

Figure 11.1 Sectoral contribution to NSDP in Nagaland

Source: Appendix A11.1.

the expenditure incurred to counter them. In absence of such detailed information, an attempt has been made to highlight the performance of economic sectors (contributions to NSDP) by estimating their annual exponential growth rates under different periods. This gives an overview of the growth of the state's economy under the armed conflicts.

The period under consideration is from 1981 to 2006. Due to problem with the data, other periods are not included. During this period, 1981–90s, the situation was volatile especially during the 1990s as compared to 2000 onward because of ceasefires between the conflicting groups, and the situation had turned relatively peaceful.

The figures in Table 11.1 and also in Appendix A11.2 indicate that during 1980–81 to 2005–06 the annual growth rate of NSDP and the per capita NSDP were 15.72 per cent and 10.95 per cent, respectively in the state. The inter-temporal growth rate of NSDP shows a declining trend from 15.78 to 14.09 per cent and 07.25 per cent per annum during the 1980s, 90s, and 2000s, respectively. Among the sectors in NSDP, the fastest growing sector was primary with 16.09 per cent per annum. It was followed by tertiary and secondary sectors with 15.85 per cent and 15.19 per cent per annum, respectively.

Table 11.1 Annual growth rates of NSDP of Nagaland (%)

Years	*Primary*	*Secondary*	*Tertiary*	*Total NSDP*
1981–1991	13.19	19.85	15.72	15.78
1991–2001	15.32	18.97	12.31	14.09
2001–2006	8.31	5.77	7.01	7.25
1981–2006	16.09	14.62	15.85	15.72

Source: Appendix A11.2.

- **Primary sector**

The inter-temporal annual growth rates of primary sector are indicated in Table 11.1; during 1981–91 it was 13.19 per cent, which further increased to 15.32 per cent during 1991–2001, but declined to 8.31 per cent during 2001–06.

Among the primary sub-sectors, mining and quarrying have emerged as one of the prominent sub-sectors in economy with a growth rate of 19.74 per cent per annum during 2001–06; with the various developmental activities together with construction activities taking place during the relatively normal period. However, the growth rate of agriculture and fishery activities had fluctuated, showing a declining trend in their growth trajectories during the 1990s, even till 2000s as indicated in Appendix A11.2.

- **Secondary sector**

The inter-temporal annual growth rates of secondary sector are indicated in Table 11.1; during 1981–91 it was 19.85 per cent, but declined to 18.97 per cent in 1991–2001, and further declined to 5.77 per cent in 2001–06.

The growth rates of sub-sectors in Appendix A11.2 show that from 1981 to 91, construction has the highest annual growth rate with 19.63 per cent and continued to increase to 20.58 per cent during 1991–2001 but witnessed a sharp decline to 1.83 per cent during the 2000s. On the other hand, there was a sharp decline in annual growth rate of manufacturing sector during 1991–2001, where it was down to 6.33 per cent from 13.08 per cent. It may be due to the fact that the medium industries like sugar mill at Dimapur and paper mill at Tuli became non-functional during 1990s. Besides, the earnest efforts of the government to extend a number of facilities to the intending industries and entrepreneurs have failed to maintain an increasing rate of growth during the critical period. Thereafter, during 2001–06, manufacturing sector exhibited the highest annual growth rate among the sub-sectors with 18.43 per cent.

- **Tertiary sector**

The inter-temporal growth trends of tertiary sector in Table 11.1 show that during 1981–91 it had increased at a rate of 15.72 per cent per annum, but declined to 12.31 per cent and 7.01 per cent per annum during 1991–2001 and 2001–06, respectively.

Among the sub-sectors as indicated in Appendix A11.2, transport and communication exhibited the highest annual growth rate during 1981–2006 with 28.17 per cent. And inter-temporal growth trend also points relatively higher growth rates for this sub-sector. It is observed that during 1981–91, the highest growth among the tertiary sub-sectors was banking and insurance with 24.10 per cent per annum. In the subsequent periods, the sub-sectors like real estate, ownership of dwelling and business services have emerged as the dominant sub-sector that has exhibited the highest annual growth rates among the tertiary sub-sectors during 1991–2001 and 2001–06 with 16.58 per cent and 10.56 per cent, respectively. It is also observed that during the last decade (2001–06), although positive, there was a decline in the growth rates for all the sub-sectors except banking and insurance.

In general, the inter-temporal performances of economic sectors revealed by their annual growth trends exhibited higher growth rates during the 1980s but decline in the subsequent decade when the political situation became more volatile, having its spill over through the 2000s, when the situation was normalised with the declaration of ceasefire, the growth rates do not show improvements but rather slowed down.

No doubt, armed conflicts effect the economic growth negatively, but it is not the sole reason for poor performance of the economic sectors in the state. Despite huge fund flow from the centre for development and its potential natural resources, the economy of the state has been at backwater mainly due to inadequate infrastructure, poor connectivity and rent-seeking attitude.

Rent-seeking attitudes

Over the years, the rent-seeking attitudes have become extensive including the insurgents. This hampers the development programmes and projects. Many who amassed huge public funds for themselves spent on such activities as luxuries and real estate that adds very little to the productive capacity of the state's economy but retards development and welfare of the general public.

The extortion and unabated tax at exorbitant rate collected by various insurgent factions from the business units discourages private investment and the growth of entrepreneurs in one hand; and on the other, the public ultimately pay accelerated prices on the essential commodities, reducing their welfare. Besides, the state government employees have been paying annual deduction arbitrarily imposed by the insurgents. On all these, the state government machinery has no control or rather turn a blind eye. For instance, for each truck load of commodities entering Nagaland, almost 50 per cent of the tax is imposed by the insurgents; other half is paid to the state government. The huge amount of resources collected from such activities is presumably spent on their organisation's upkeeps and for their personal gains.

The hardship of the general public and their open resentment of such practices had led to public upsurge against the unabated taxation in 2013 under the banner the Action Committee against Unabated Taxation (ACAUT). Further, the Business Association of Nagas (BAN) has appealed to the government 'to protect the citizen rather than be mute spectators to the sufferings of the public' (*Nagaland Post*, 2014).

The impacts of insurgency on economic growth and development are (i) resources drain by the insurgent factions, (ii) diversion of resources by the government to maintain the military troops (counter-insurgency), (iii) shaken business confidence and risky investment, (iv) perpetual flight of capital through institutional mechanism resulting in low credit deposit ratio of banks, (v) unproductive public expenditure in the interest of power elites to maintain their power base and (vi) inflated prices due to multiple taxes; ultimately in all cases welfare of the general people is reduced.

Conclusion

Nagaland has been experiencing disturbing and painful developments with the advent of the British authority in its territories, followed by the independence of India. The Indo-Naga political conflict of more than half a century has jeopardised the process of development and reduced human welfare in the state. The issue remained unresolved till today despite having a number of peace talks and ceasefires between the Government of India and the insurgent factions. The present ceasefire and peace talk is reported to be a routine process without any concrete solution. Thus, the uncertainty looms high for the Nagas.

It may be summarised that Nagaland has been experiencing a situation with the proliferation of arms as well as activities of insurgent factions in one hand and the military and paramilitary troops as counter-insurgency agency on the other, leaving no space for public to play any role. Moreover, the absence of business-initiative environment and administrative legal framework under parallel governments has kept the private sector away from any worthwhile investment in the state. At the same time, the power elites with rent-seeking behaviour have crippled the public sector organizations and initiatives as well.

Although the Naga insurgency movement emerged not due to lack of economic progress but in the process, they reinforced each other. Therefore, the policy-planners need to understand this relationship. It is not so easy or simple to suggest any measure to resolve this issue; nevertheless, it requires political will and sincerity on the part of both the parties to solve the issue in the right perspective, finding an honourable settlement that takes care of the people's aspirations.

For a region, its economic progress is greatly influenced by its history, sociopolitical conditions and natural factors. Nagaland, with a history of more than 50 years of armed conflicts and situated at difficult terrain, has left the land with weak industrial sector and an underdeveloped agriculture; bulk of educated population depends on service sector which is predominantly of government services, and a poorly developed infrastructure set up. Although the government sector employs a bulk of educated labour force, this sector has its limitation in creation of additional jobs. Therefore, there is urgent need for creation of environment conducive for progress of local entrepreneurship and attract private investment. In addition, agricultural sector should move towards commercialisation. All these will require strong infrastructural and institutional supports. To usher in these changes in its economic structure, the state requires to create business-initiative environment and strong administrative legal frame with a clear vision for progress and to safeguard the welfare of the masses.

References

Anand, V.K. (1980). *Conflict in Nagaland: A Study of Insurgency and Counter Insurgency*, Delhi: Chanakya Publications, p. 55.

Bhatt, C.S. and Gobal K. Bhargava (eds) (2006). *Land and People of Indian States and Union Territories, Nagaland*, Vol. 20, Delhi: Kalpaz Publications, p. 14, 18.

Chatterjee, S.K. (1950). 'Kirata Jana Krit', *Journal of Asiatic Society of Bengal*, XVI(2): 145.

Devi, Lakshmi (1968). *Ahom Tribal Relations*, Guwahati: Assam Book Depot, pp. 15–20.
Government of Nagaland (1973, 2009, 2012). *Statistical Handbooks of Nagaland*, Kohima: Directorate of Economic and Statistics.
——— (2004). *Nagaland State Human Development Report*, Kohima: Department of Planning and Coordination.
——— (2014). *Mokokchung District Human Development Report*, Kohima: Department of Planning and Coordination.
Horam, M. (1975). *Naga Polity*, Delhi: B.R. Publishing Corporation, p. 46.
Mackenzie, A. (1969). *History of the Relations of the Government with the Hill Tribes of the North East Frontier of Bengal, Calcutta*: Home Department Press, p. 102.
Nag, Sajal (2013), Expanding Imaginations: Theory and Praxis of Naga Nation Making in post-Colonial India, In Tanweer Fazal (ed), *Minority and Nationalisms in South Asia: South Asian History and Culture*, London and New York: Routledge, p. 24.
Nagaland Post (2014). 'BAN Appeals NPGs, Govt. on Taxation Demands'. 2 February 2014, Vol. 24, No. 56, p. 1.
Nibedon, Nirmal (1988). *The Ethnic Explosion*, New Delhi: Lancer Publishers.
The Indian Express (2012). '4 Eastern Nagaland Districts Seek Statehood', 3 April 2012, indianexpress.com/news/4-eastern-nagaland-districts-seek-statehood/932012/.
Wikipedia (2014). 'Shillong Accord of 1975', http:blog.wikimedia.org/2014/06/16.
Zhimomi, Kuhoi K. (2004). *Politics and Militancy in Nagaland*, New Delhi: Deep & Deep Publications.

Appendix

Appendix A11.1 NSDP at factor cost in Nagaland (Rs in Lakhs)

Sectors/ sub-sectors	*1980–81*	*1990–91*	*2000–01*	*2005–06*
1. Primary sector	**3,428*(32.5)***	**6,954*(29.22)***	**75,147*(33.69)***	**159,162*(34.60)***
Agriculture	3,022*(28.6)*	4,411*(18.53)*	67,606*(30.31)*	142,840*(31.28)*
Forestry and logging	401*(3.80)*	2,389*(10.03)*	6,307*(12.82)*	13,930*(3.05)*
Fishery	5*(0.04)*	154*(0.64)*	1,145*(0.51)*	2,068*(0.45)*
Mining and quarrying	0*(0.00)*	0*(0.00)*	89*(0.03)*	324*(0.07)*
2. Secondary sector	**1,490*(14.12)***	**6,117*(25.70)***	**31,551*(14.14)***	**63,665*(13.94)***
Manufacturing	226*(2.14)*	908*(3.81)*	1,456*(0.65)*	6,752*(1.48)*
Construction	593*(15.10)*	5,284*(22.20)*	32,303*(14.48)*	53,538*(11.73)*
Electricity, water supply, gas	329*(-3.11)*	75*(-0.31)*	2,208*(-0.98)*	3,375*(0.74)*
3. Tertiary sector	**5,629*(53.37)***	**10,727*(45.07)***	**116,344*(52.16)***	**233,781*(51.20)***
Transport, storage and communication	177*(4.67)*	1,383*(3.81)*	38,362*(17.19)*	60,846*(13.33)*
Trade, hotel and restaurant	724*(6.86)*	1,230*(5.17)*	11,308*(5.06)*	23,413*(5.13)*
Banking and insurance	141*(1.33)*	806*(3.38)*	2,761*(1.23)*	7,096*(1.55)*
Real estate, ownership of dwelling and business service	1,535*(14.55)*	1,738*(7.30)*	21,609*(9.68)*	51,882*(11.36)*
Public administration	1,952*(18.50)*	2,500*(10.50)*	27,017*(12.11)*	54,880*(12.02)*
Other services	1,100*(10.42)*	3,070*(12.90)*	24,051*(6.85)*	35,664*(7.81)*
NSDP	**10,547**	**23,798**	**223,042**	**456,608**
Per capita income (Rs)	**1,361**	**2,051**	**11,473**	**18,318**

Source: Statistical Handbooks of Nagaland 1983, 1986, 1991, 1996, 2000, 2004 and 2009. Directorate of Economic and Statistics, Nagaland, Kohima

Notes

1. The figures in brackets represent the percentage shares in Net Domestic Product.
2. Based on 1980 prices for 1980–81 and 1990–91 and 1999–2000 prices for 2000–01 and 2005–06.

Appendix A11.2 Sector-wise annual average growth rates of NSDP in Nagaland

Sector	*1980–81 to 2005–06*	*1980–81 to 1990–91*	*199–91 to 2000–01*	*2000–01 to 2005–06*
1. Primary sector	**16.09 (.988)**	**13.19 (.886)**	**15.32 (.970)**	**08.31 (.989)**
Agriculture	16.78 (.983)	11.12 (.807)	15.78 (.978)	08.77 (.975)
Forestry and logging	11.89 (.919)	20.62 (.908)	11.45 (.735)	05.24 (.539)
Fishery	19.25 (.891)	31.57 (.725)	25.91 (.978)	01.62 (.054)
Mining and quarrying	00	00	00	19.74 (.795)
2. Secondary sector	**14.62 (.954)**	**19.85 (.950)**	**18.97 (.815)**	**05.77 (.592)**
Manufacturing	12.49 (.701)	13.08 (.705)	06.93 (.111)	18.43 (.920)
Construction	13.78 (.939)	19.63 (.916)	20.58 (.927)	01.83 (.100)
Electricity, water supply and gas	-	-	-	07.02 (.848)
3. Tertiary sector	**15.85 (.971)**	**15.72 (.985)**	**12.31 (.960)**	**07.01 (.956)**
Transport, storage and communication	28.17 (.943)	22.49 (.831)	28.36 (.811)	08.00 (.858)
Trade, hotel and restaurant	15.38 (.970)	13.69 (.957)	12.77 (.952)	04.33 (.938)
Banking and insurance	14.01 (.549)	24.10 (.901)	−19.29 (.226)	08.31 (.934)
Real estate, ownership of dwelling and business service	15.71 (.978)	11.11 (.974)	16.58 (.926)	10.56 (.928)
Public administration	13.69 (.972)	14.99 (.977)	12.64 (.944)	03.62 (.687)
Other services	13.22 (.953)	19.02 (.982)	07.23 (.921)	07.56 (.948)
NSDP	15.72 (.984)	15.78 (.983)	14.09 (.956)	07.25 (.992)
Per capita income	10.95 (.964)	12.32 (.972)	09.13 (.902)	03.09 (.934)

Source: Statistical Handbooks of Nagaland 1983, 1986, 1991, 1996, 2000, 2004 and 2009. Directorate of Economic and Statistics, Nagaland, Kohima.

Note: The figures in brackets represent r^2 (coefficient of determination).

Chapter 12

A nation's begotten child

Arunachal Pradesh in India's troubled Northeast

Tajen Dabi

Introduction

Arunachal Pradesh, earlier known as *North East Frontier Agency* (NEFA), attracted the attention of post-independence Indian state and political analysts after the Chinese invasion of 1962 in a major may. This chapter draws a general outline of the recent history of Arunachal Pradesh and tries to understand why the state is 'peaceful' in the otherwise politically restive Northeast India. The emergence of ethno-religious identity is linked to the larger process of 'nation building' and modernisation and shows that ethnopolitical consciousness in the state is, unlike most other parts of the region, taking roots in manners that does not contradict with the terms of 'mainstream' ideology-cultural and political. This chapter tries to develop a critique of the post-Nehru policy of Central Government and some other non-state agencies in Arunachal Pradesh.

Island of peace in Northeast India

Post-independence India witnessed numerous secessionist movements. Northeast India is known more for such movements. Nagas, Mizos, Meiteis, Assamese, Tripuris, Bodos, Kukis and Karbis, in random order, among others, have different ideological roots, inspirations and methods in pursuance to their claims and 'national' struggle. In this context, Burman (2007: 24–40) writes:

> Leaving aside the Bengali population of Barak and Brahmaputra valleys, and taking stock of the overall historic-political situation it can be said that while in general the peoples of the North-East have mixed feelings about their place in the state structure of India and the articulation of the same with the state processes of the country,

> they frequently entertain mytho-poetic sentiments about their ties with South-East and East Asia.

Despite frequent fights among them, the communities were all opposed to the Indian state (Guha, 2007: 625). Their struggle were, and are sometimes directed against each other's interest, the more known and recent one being the Bodo movement (presumed by the 'mainstream' Assamese political and intellectual group as an attempt to 'divide' Assam). Such movements are absent in Arunachal Pradesh. It is commended as the 'island of peace' – in academia and in public – amidst her restive neighbours. This view reflects a law and order perspective bias in the assessment of ethnic situation in the region in general and Arunachal Pradesh in particular.

The context of 'peace' in Arunachal Pradesh

The Northeast India[1] constitutes two geographical regions: plains (Assam – Brahmaputra and Barak valley) and the Hills (roughly starting from south – Mizoram, Tripura, Manipur, Meghalaya, Nagaland and Arunachal Pradesh). As per historical records, the main highlights of the relation between the state(s)/kingdom(s) of the plains and the roaming semi-pastoralist groups/chiefdom(s) of the hills were – one, barter economy, and two, raids from the hills into the plains. While the barter trade kept lines of communication open within the hills and between the hills and the plain, the raids into more affluent plains suggest that the trade did not evolve in a regular, profitable and sustainable manner. Throughout much of Ahom and British colonial times, the nuisance of raids[2] remained a pressing political question of the day, and at times guided strategic policies of the respective governments. It would be no exaggeration to say that the *Inner Line Regulation*,[3] the most important and, to some, controversial piece of law introduced by the British that allegedly 'created' and continues to separate the hills from/and the plains, was enacted as an institutional response to such raids. That the context and meaning of the regulation have undergone much change, now acquiring cultural–territorial dimension, is only indicative of the fractious notions of sovereignty each evolving identity groups harbour – in the hills and the plains, against each other, and against India. Thus, political contestations and disturbances, in its own specific context and relevance, were as much a reality in the past as is in the present. The recent history of what we call 'Arunachal Pradesh' today did not emerge from a situation significantly different from this. Historical, archaeological, ethnographical and folk traditions and memory attest to this.

Assuming the various autonomous-secessionist movements as (i) either a hangover of the old values or (ii) a mere continuation of raids, now directed, in expanded and extreme(ist) tone, against the power that replaced Ahom Kingdom and British, how do one rationalise the 'peace' of Arunachal Pradesh? One ethnic group or community – presumed or real – after another rebelled, and is rebelling against the Indian state, and in many cases against the supposed or factual domination of one by another within the region. How, then, does Arunachal Pradesh come to acquire a personality completely different from others?

Except Arunachal Pradesh, the other states of the region are either cultural successor(s) of (a) preceding state system(s) or chiefdom(s) or a part of it. At the time of the annexation of the Northeast region by the British, there were *six major tribal states under the Hinduised rulers* – Koch, Tripuri, Jayantiya, Kachari, Ahom and Meitei (Bhadra, 2007: 2). Does this historical backdrop offer us some insights into the recent history and politics of the region, and the state in particular? Does the relative 'backwardness' of communities in Arunachal Pradesh in the evolutionary graph confine them in a 'peaceful' sociopolitical ladder in the otherwise contested national identities of, at least the way the 'sub nationalists' (Baruah, 1999) assert themselves, the contemporary northeast? Is the relative peace of Arunachal because of the inherent structural inadequacies, as a recent research in the case of Mishings (as compared to Bodos) suggested (Saikia, 2011)? If it is so, how then, do the embryonic political and intellectual classes of the state represent and assert themselves? Does it take a proud common 'Arunachali' stance; even make pretence to it, or does it take a deeper, sectarian stance where one can visualise ethnopolitical violence/movements so common in nearby Assam? What factors and circumstances accompanied the quick and seemingly smooth journey this hinterland made, quite exceptional when compared to other states of northeast, in becoming a 'peaceful', non-rebellious state of Northeast India? What is the role of post-independent government policy? What image does the word 'Arunachal Pradesh' conveys? Is there an 'Arunachali' the way we have 'Nagas', 'Assamese', 'Bodos', 'Mizos', 'Meiteis' and so on capable of acquiring a rebellious reputation that has become so emblematic of the region now?

Historical backdrop: the sources and their employers

We can list, in random order, the following categories of historical sources for Arunachal Pradesh extant: First, popular religious literature of the mainstream Hindu tradition and second, the 'foreign' accounts.

In both, there are confusing references to the region, its geography and inhabitants. The generic term *Kirata* is a classic example of how stray, indicative and uncorroborated references are employed to suit retrospective identities and cultural agenda. 'Kirata' is a holy grail to establish the antiquity of various hill people, and their participation in the epic battle at Kurukshetra. Such academic exercise resonates with Sanjib Baruah's (1999: 180) accusation that

> The history of the assimilation of the Bodos and of many 'tribal' peoples into the Assamese formation provides one of the most dramatic examples of how Indic civilization in India's Northeastern periphery managed to recruit converts from the supposed "primitive" peoples, and of the continuity between caste and the supposed "primitive isolates" – the "tribes".

The next set is the literature of the Ahom period, notably the *Buranjis*,[4] among others. Fourth, the various records of the colonial period, and fifth the more recent research monographs and travelogues.

The approach to preliterate societies like those of Arunachal Pradesh, in both academia and state policy, has been with a sense of burden and patronage. Colonial literature is a convenient kick bag to ventilate ones' aversion to such attitudes. Even the towering historian of region, Barpujari's masterpiece, *Problem of the Hill Tribes North-East Frontier*, has been accused, respective or irrespective of the context, of viewing the tribal question as a 'problem' (Hilaly, 2008: 415–9). Many have questioned the use of conventional models and generalisations and urged the need for alternative schemes to write the history of the region (Momin, 2003: 32–44; Showren, 2006: 46–54; Syiemlieh, 2010: 1–15).

The employment of historical sources has intimate connection with the question of ethnic assertion and 'subnationalistic' politics in Northeast, and how these are contested by the 'mainstream' ideology. Most insurgent groups and their intellectual think tanks cite 'historical evidence' for their separate political and cultural entity vis-à-vis India. Respective groups and even local historians routinely trace the 'sovereignty' of Assam and Manipur back to the Treaty of Yandaboo of 1826 (Goswami, 2012: 17). Naga rebels go back to the 1940s to claim their 'prior independence' before India became one. Bodos, fighting against both Assam and India, relive the memory of old pre-Ahom Kachari kingdom in their Bodoland struggle, amply reflected in the proposed territorial jurisdiction of 'Greater Bodoland'.

The silence of the lamb[5]

The subnationalistic baptism into past along with its attendant political stance is absent in Arunachal Pradesh. It is a result of two sets of conditions that evolved into an axis whose direction was smoothly set in the tenor of post-Independent India's nation-building process. The first set of condition is the greater degree of isolation that Arunachal Pradesh experienced, in its entire history prior to its incorporation into Indian state, compared to other states of the region. Trade relations with neighbouring Tibet did not result into larger political and cultural ties (with the exception of Mahayana Buddhist areas in Tawang, Dirang, upper reaches of Subansiri, Siang and Dibang), and relations with the Brahmaputra valley mainly revolved around one singular aspect, that of the *Posa*.[6] Debates on its relevance apart, probably no other instrument of law concretised and reflect this isolation than the *Inner Line Regulation*.

As a result, some of the important factors for intellectual awakening and political mobilisation in the context of 'tribal Northeast' during the British period – modern education and Christianity did not grow in Arunachal Pradesh. Lying between the loosely defined *Inner Line* and the vague 'Outer Line',[7] it never engaged the mind of the colonial state, well consolidated in Brahmaputra valley by mid-nineteenth century, beyond imperial strategic exigencies and sometimes trade (Sikdar, 1981: 210–20). It was, in official parlance, a frontier that graduated from a 'Tract' (*North East Frontier Tract*) to an 'Agency' (*North East Frontier Agency, NEFA*) then as a Union Territory[8] (in 1972) and finally as a state (in 1987) of the Republic of India.[9]

The second set of condition is a peculiar mix of varied traditions and how it impeded evolution of a cohesive social order. Verrier Elwin[10] has identified three cultural groups in the state viz. one, those who are practicing or influenced by Buddhism; two, the Tani group of tribes, the Mishmis and 'sundry tribes' in the central zone and third, the Noctes, Tangsas and the Wanchos in the east (Bose, 1997: 22). The first group, professing lamaistic Buddhism (Nyigmapa and Gelukpa sects), had cultural and loose political link with Tibet until recent times (Haimendorf, 1982: 148–9). Their social and political life centered on their religion. The tribes of the central zone practiced animistic religion. Social order was maintained and political actions mobilised either by or through a combination of a network of extended household, family, clan or village as the primary unit. There was practically no instrument of governance above these units. Haimendorf, writing decades after Elwin, describes some of the Subansiri tribes as 'Lawless society' (ibid.: 98). Were we to turn the wheel of time some more decades back, the description will be well apt for other tribes of this zone – Akas (Hrussos), the Mijis (Sajolangs), Apatanis, Tagins, Galos, the Adis, the Mishmis and others. In other words, there was no political 'tribe'. In the midst of 'high and rugged' mountains, the people were kept 'divided' (Scott, 2009: 16).

The third group – Noctes, Tangsas and the Wanchos – maintained a quasi-oligarchic polity with a marked hereditary institution of Chieftainship (Elwin, 1965). It is this group of tribes and the area they inhabit that the Nationalist Socialist Council of Nagaland is now claiming to be 'unified' under the proposed 'Greater Nagalim'. The Scheduled Tribe Order lists these tribes as Naga 'subtribes'. The Khamptis and Singphos are the other tribes in the eastern zone. Both practice Hinayana Buddhism (some section of the latter follow Vaishnavism, a creed popular in plains in neighbouring Assam), share close cultural and linguistic ties with the Tai-Ahoms and their eastern neighbours in Myanmar and South East Asia (Burman, 2007: 27). Thai movies are popular among Khampti youths and village folks and the latter claim to 'understand 80 per cent' of the language spoken in these movies.[11]

These communities of the state exhibit considerable degree of diversity in terms of religion, language and customary practice. Where cultural and linguistic ties exist – such as in the case of most of the tribes of central zone – conditions for growth of a cohesive social and political order were absent until present times. That is, the process of formation of 'tribes' – the basis for ethnic identity assertion – began *only* in the wake of modernisation induced by post-independent nation-state. From the Elwin–Nehru protectionist – 'isolationist' approach of the 1950s – the direction of the axis I spoke about earlier started to acquire a 'nationalistic' turn after the Chinese aggression in 1962. Along with rapid administrative changes, budgetary allocation from New Delhi began to increase. Writes a commentator in the early 1980s, 'Within a 20 year time span, the emergence of new economic basis has given the appearance of a social revolution taking place in Arunachal . . . no outsider worth the name are involved in this transformation . . . except through the agency of the state' (Mishra, 1983: 1837). No wonder then, that when made a Union Territory in 1972, it was given a Sanskrit-Hindi name – 'Arunachal Pradesh'. A song popular in much of central Arunachal (where there is linguistic similarity, a fact that was, probably first, noted by M. Robinson in 1851 and echoed by both Elwin and Haimendorf a century later), frequently sung during VVIP[12] visits from Delhi, aptly represents the supercilious[13] metaphor of Arunachal Pradesh as the 'land of rising sun' of India:

> *Aro ge kamchi ge Donyi ge*
> *Tupa choko nguluge Arunachal!*
> *(Where the Sun rises first at the break of dawn*
> *Our land, Our Arunachal!)*[14]

This 'nation building' is reflected in another instance of popular culture. In 1976, a film was released, quite an innovation for the time, provocatively titled in Hindi, *Mera Dharam Meri Maa ['My Duty (religion?) My Mother (motherland – India?)']*, produced by the Government of Arunachal Pradesh and directed by the ace singer and filmmaker from neighbouring Assam, Dr Bhupen Hazarika. It is set in the backdrop of transition into 'modern life' – school, medical facilities, student politics and so on. The film is didactic in intent. Characters are drawn from local tribals who speak in Hindi and use innuendoes common to Bollywood movies – hardly indicative of tribal ethos. In one of the important plot of the movie, a local boy is reminded that he is a 'Suryaputra' ('Son of Sun God', a concept popular in Hindu religion) in order to provoke his suppressed sense of masculinity.

The chief objectives of state-sponsored modernisation were three-fold: strategic, economic–social and cultural. Since the British colonial period, the area has been viewed more in strategic terms, and continues to be so. Socio-economic concerns such as food, medicine, road communication and physical infrastructure were such benefits deservedly extended to the tribes since the advent of Indian administration (Elwin, 1957). The cultural concerns largely tended to drift away from Elwin's 'cultural aims for NEFA' (ibid.). This can be seen in two important sectors: education and religion – probably the most important tools of social engineering in a preliterate and largely animistic society. Hindi (replacing Assamese), along with English, began to be taught in schools. No effort was made at institutional level to encourage learning of indigenous languages. Where Christian missionaries developed scripts and dictionaries in many languages in other parts of region, which latter on became a basis for cultural regeneration, similar efforts are initiated at the mercy of fledging community-based literary societies in Arunachal today.

Decades after modern education was started, religion assumed prominence in public policy. With an objective to stop conversion to Christianity, the Arunachal Pradesh Freedom of Religion Act, 1978, officially inaugurated the most challenging and controversial, social question – religious conversion. The Act recognised Buddhism, Vaishnavism and Nature worship as 'Indigenous faith'. That conversion to Christianity was a growing threat to indigenous faith is not debatable. Since 1950s, Christianity has been spreading in the state, and unlike other parts of Northeast, its spread in the state was a result of social instability in the face of rapid modernisation. The new faith attracted converts mostly from among the 'Nature worshipers'. In the absence of a corresponding class of intellectuals, the 'petty bourgeoisie and the bourgeoisie' of the nineteenth century – India reform movement – the question of 'the destruction or denigration of indigenous culture' (Panikkar, 2001: 79) in Arunachal Pradesh offered *space to external agencies to appropriate, formulate and direct the nature of response* to such challenges. And such response often took quite a circuitous path, mostly pursuing a discreet 'national' agenda in the garb of 'protecting' the indigenous culture.

It is under such circumstances that the native response to the threat of indigenous culture has taken place, and still is. The most glaring example of this is the way the indigenous religion itself has come to project itself. Among the 'indigenous faiths' mentioned in the Arunachal Pradesh Freedom of the Religion Act of 1978, the basic concept of the 'Nature worshipers' is getting fundamentally transformed. The 'New Dispensation', to borrow the appellation given to Keshab Chandra Sen's reformed Brahmo Samaj (ibid.: 28) adopted *prayer*, *congregation*, *church* and *gods* to its reformed religion; nothing remained 'indigenous' except the growing personification of animistic deities into 'gods' and 'goddesses' – in the form of idols. These gods share divine platform alongside Hindu gods in the reformed churches – of appropriate denominations, viz., Donyi-Polo*ism*, Nani-Intaya*ism*, Rangfra*ism*, maintained by various quasi-religious community organisations throughout the state.

The earliest exponent and leadership of this reformist movement was provided by now revered thinker from Siang (one of the earliest centre of modern education in Arunachal Pradesh), late Mr Talom Rukbo. The social base of the leadership of this movement consists of first generation of educated people who are well placed in government jobs and their retired counterparts; aspiring or retired (but never a serving) politician; folk rhapsodist and so on. To their rank, they recruit the traditional ritual performers (shamans), who *nonetheless continue to perform the animistic rituals in their own private capacity* whenever asked so, *a practice that has no institutional (and theological connection with the reformed religion.* To consider that their intention is malicious would be a wrong assumption any more than one can presume Rammohan Roy to be an agent of the missionaries. It is their connection to external agencies, and the guidance/patronage they derive from it, that betrays the roots of the movement's non-indigenous character.

Local dailies routinely update collaborative efforts of the *Indigenous Faith and Cultural Society of Arunachal Pradesh* (IFCSAP) and radical Hindu organisations. With the chief objective of checking conversion to Christianity, this reformist movement has so far achieved nothing more than to bridge a tenuous connection with Hinduism, thereby proving Roy Burman right that, 'many, even in the hills of the North-East certainly try hard to ingress India in their existential commitment' (2007: 24). As if to lay claim to the success of these guided reformist movement, well-known Yoga-guru Ramdev in his maiden visit to the state controversially said, 'Every follower of indigenous faith is a Hindu' (*Sentinel Arunachal*, 17 February 2011). The President of the IFCSAP, a litterateur in his own right, taking wrong lessons from history, or right lessons from wronged history, confirmed the visiting seer saying, 'History tells (*sic*) Arunachal Pradesh is a Hindu state' (*Arunachal Times*, 17 February 2011). A process of, to borrow and corrupt a phrase, *textualised discovery* (Dirks, 2010: 297) of *oral* indigenous faiths is in order.

The predicament

What is the future of Arunachal Pradesh – in ethnic, religious and political terms – in itself and in context of the political situation in the region? The above-mentioned religious-reformist movement does not address this attendant political question; it conveniently avoids it. That is, the cultural link sought to be established between indigenous faiths and Hinduism does not address the social and political questions affecting the future of the indigenous communities; the religious question ignores and negates any possible use of its shared values, network and organisation to assuage rising political anger and discontent in the state.

The rising demand for 'autonomy' is the barometer of this dormant ethnopolitical discontent. The state government has referred two such bodies – the Patkai and Mon Autonomous Council (roughly corresponding to Elwin's eastern and western zone – the 'Naga' and Buddhist regions, respectively) to New Delhi. Will the setting up of these autonomous councils, if it becomes a reality, trigger demands for more of such autonomous political spaces? It is very likely. Because both the above two regions correspond to an assumed 'ethnic' and cultural space, distinct from others. It will give impetus to the belief that 'tribe' is a convincing category to augment presumed economic, social and political aspirations – justified or otherwise. If such is the political future of the state, how are matters of common interest then attended to? Dams, illegal migration and massive military presence are matters of concern in the region (Guha, 2007: 626–27). The first two problems acutely affect the future of the indigenous people of Arunachal Pradesh also. Add to this the frequent clash-over-boundary dispute with neighbouring Assam, which the indigenous people, losing their ancestral lands (Haimendorf, 1982: 27) by day, believe that both the Assam and Central government turn a blind eye in order to 'adjust' the illegal migrants from Bangladesh into thinly populated foothill areas of the state. The Chakma-Hajongs, Buddhist political refugees from Bangladesh, were settled in parts of the state with the same logic; the issue is another current political question in the state.

The repeated clarification (and reassurance to pretentiously patriotic local politicians) by New Delhi that Arunachal is an integral part of India and that its territorial integrity will never be compromised, is a fact no one questions or is genuinely bothered in Arunachal. While the Chinese claim on Arunachal is a window to staged patriotism and 'national' politics for local political class, it has become a convenient tool to augment broad 'nationalistic' policy in Arunachal Pradesh better than in any other states of the region. Language (Hindi), demography (legal and illegal migration), economy (dams, mining) and culture (Hinduism) are slowly but steadily taking a 'national' turn while the local social and intellectual capital is wasted in pursuing ends that either complement these or sharpen growing ethnic divide.

These throw up a seemingly historical irony in the Northeast: Naga-Kuki-Mizo-Khasi and many other tribes employ Christianity to assert their separate 'nationality'/identity. The process is opposite in Arunachal, *rejection* of Christianity is one vital aspect to identify itself with the Indian nation. The Meiteis, sections of Tai-Ahoms and many others, are reasserting their 'tribal' identity now as opposed to Hinduised identity. Similarly, religious reform in Arunachal is towards seeking a place within Hinduism. Many local scholars even draw analogous functions with Hindu gods, and goddesses, to describe their animistic gods. Where Shillong, one of the earliest and best-developed administrative centres by the British in the region, once capital of 'united' Assam, is burning today demanding for the *Inner Line Regulation* to be introduced in Meghalaya, railways is being introduced in Itanagar, the state capital. It is ironical that while both the Christianised and Hinduised communities of region try to project a separate nationality vis-à-vis the Indian state, the case is different in Arunachal Pradesh. Others in the region try to rediscover and relive the time before the Hindu priests found them and British Empire defined their political future, Arunachal is a pampered boy growing into the manhood of Indian nation.

Conclusion

In assessing the state of politics and development in Arunachal Pradesh, especially in contrast to its neighbours in the region, the historical roots of its isolation and the way it was integrated into Indian administration are a matter of great importance. Both historical antecedents and post-independent Indian state's concerns coalesce into a complex set of processes that characterises the face of politics on, and within Arunachal Pradesh today. In contrast to most parts of the region, identity politics in Arunachal, as of today, does not draw a distinction with the metaphor, symbol and idioms of Indian nation and instead forges a link with it.

At the surface, the symbolic pillars of 'separateness' viz. *Inner Line Regulation* continues to exist but fundamental changes are being effected subtly through rapid modernisation and developmental projects. The absence of a militant ideology, a feature so common in the region, directed against the Indian state is probably because of two reasons. One, the sense of deprivation and alienation, if any, is yet to take an ideological dimension and/or two, the nature of integration of the communities (and their neo-political elites) into the narrative of Indian nation has been comparatively more accommodative than in other parts of the region. Development and ethnic politics in Arunachal Pradesh is closely connected with the long-term strategic, economic and cultural course of the Indian nation state. It has been an active 'collaborator' in the process of structural changes in the societies of the indigenous communities of Arunachal Pradesh. In addition, unlike in other parts of the 'schizophrenically alienated' (Burman, 2007: 24) Northeast India, it has successfully, so far, mediated the modernisation of the tribes of Arunachal Pradesh in line with the larger path of the Indian nation state, even if it was incongruent with Elwin's philosophy and Nehru's foresight.

Notes

1 Excluding Sikkim.
2 'Rapine and lawless aggression on their lowland neighbors', in the words of M. Robinson (1851).
3 Introduced by the British in the last decades of nineteenth century, the line demarcated the operational limits of the regular administration. British subjects were not allowed to go without official permission, neither

own land beyond the line. It is still in force in Nagaland, Mizoram and Arunachal Pradesh. Various civil society organisations in Meghalaya are now demanding for the same to be introduced in the state.

4 Family chronicles maintained by the Ahom kings.

5 Borrowed from the title of Thomas Harris's 1988 novel, *The Silence of the Lambs.*

6 Tax/blackmail/ tributes levied by various, but not all, tribes on the neighbouring peasants and gold washers of plains. The latter were subjects of Ahom Kingdom, and later the Colonial state. Raids were organised chiefly for extracting *Posa* which was given in kind (Ahom time) and cash (British period). Interestingly, *Posa* is another arena where the presumed 'supremacy' is contested now both by historians based in Plains and Hills. This assumes significance in the light of present interstate boundary disputes Assam (plains) has with its neighbours – Nagaland, Meghalaya and Arunachal Pradesh, roughly all former Posa-receiving/ exacting parties.

7 De facto imaginary line between British India and Tibet conceptualised in the last decades, a time when British India was hardly bothered about the international border with China, of nineteenth century. The more precise, and controversial, Mc Mohan Line demarcating the international boundary came up in the second decade of twentieth century.

8 Constituent part of the Indian union directly administered by the Union government.

9 Until after Chinese aggression (1962), it remained under the Ministry of External Affairs, Government of India.

10 Noted anthropologist, tribal sympathiser and adviser to the Government of India (Nehru) on tribal issues.

11 Interview with Khampti students, Rajiv Gandhi University, Itanagar.

12 *Very Very Important Person*; a part of Indian political culture much aped in Arunachal.

13 One of the popular songs of the noted Assamese singer, Dr Bhupen Hazarika also proudly asserts Assam as the 'the first province of Bharat where the sun first rises'. It is interesting to note that Assamese political leaders and intellectuals of last generation have a nostalgia with the name 'NEFA', the erstwhile name of Arunachal and during which time they assumed NEFA to be a 'part' of Assam. 'Arunachal' signifies the symbolic break of ties – political and cultural – with 'Assam's NEFA'.

14 Credited to a local singer from Upper Subansiri, Takio Soki, the song has since acquired many versions; translation author's.

References

Barpujari, H.K. (1981). *Problem of the Hill Tribes North-East Frontier: 1843–72*, 3 vols. Guwahati: Lawyers Book Stall.

Baruah, Sanjib (1999). *India against Itself: Assam and the Politics of Nationality.* New Delhi: Oxford University Publication.

Bhadra, R.K. and Mita Bhadra (Ed.) (2007). *Ethnicity, Movement and Social Structure: Contested Cultural Identity.* New Delhi: Rawat Publication.

Bose, M.L. (1997). *History of Arunachal Pradesh.* New Delhi: Concept Publications.

Burman, B.K. Roy (2007). 'For Overcoming the Schizophrenic Alienation of the North-East: Outline of a Comparative Approach', In R.K. Bhadra and Mita Bhadra (Ed.), *Ethnicity, Movement and Social Structure: Contested Cultural Identity.* New Delhi: Rawat Publication.

Dirks, Nicholas B. (2010). *Castes of Mind: Colonialism and the Making of Modern India.* Delhi: Permanent Black.

Elwin, Verrier (1957). *A Philosophy for NEFA.* 5th Reprint 2006. Itanagar: Directorate of Research, Government of Arunachal Pradesh.

——— (1958). *Myths of the North East Frontier of India.* 2nd Reprint 1993. Itanagar: Directorate of Research, Government of Arunachal Pradesh.

——— (1965). *Democracy in NEFA.* 1st Reprint 2007. Itanagar: Directorate of research, Government of Arunachal Pradesh.

——— (1970). *A New Book of Tribal Fiction.* 1st Reprint 1991. Itanagar: Directorate of research, Government of Arunachal Pradesh.

Goswami, Priyam (2012). *The History of Assam: From Yandaboo to Partition, 1826–1947.* New Delhi: Orient Blackswan.

Guha, Ramachandra (2007). *India after Gandhi: The History of the World's Largest Democracy.* London: Picador India.

Haimendorf, Christoph von Furer (1982). *Highlanders of Arunachal Pradesh.* New Delhi: Vikas Publishing House.

Hilaly, Sarah (2008). 'Representation of the Ethnic Communities of North-East: An Overview', Proceedings of North East India History Association (NEIHA), Dibrugarh.

Mishra, S.N. (1983). 'Arunachal's Tribal Economic Formation and their Dissolution', *Economic & Political Weekly,* 18(43): 1837–46.

Momin, Mignonette (2003). 'Generalization in Constructing Histories of North East India', Proceedings of NEIHA, 24th Session, Guwahati.

Panikkar, K.N. (2001). *Culture, Ideology, Hegemony: Intellectuals and Social Consciousness in Colonial India.* London: Anthem Press.

Robinson, M. (1851). 'Notes on the Dophlas and the Peculiarities of Their Language'. Assam Commissioner's Office, F/No. 420, Assam State Archives, Dispur.

Saikia, Pahi (2011). *Ethnic Mobilisation and Violence in Northeast India.* New Delhi: Routledge.

Scott, James, C. (2009). *The Art of Not Being Governed: An Anarchist History of Upland Southeast Asia.* New Haven and London: Yale University Press.

Showren, Tana (2006). 'Ethnohistory in Arunachal Pradesh: Difficulties and Scope', Proceedings of NEIHA, 27th Session, Aizawl.

Sikdar, Sudatta (1981). *Cross-Country Trade in the Making of British Policy towards Arunachalis in the Nineteenth Century.* NEIHA: Dibrugarh.

Syiemlieh, David R. (2010). 'Presidential Address', Proceedings of NEIHA, 31st Session, Tura (Meghalaya).

Chapter 13

Democracy and ethnic politics in Sikkim

M. Amarjeet Singh and Komol Singha

Introduction

Sikkim was an independent kingdom, founded in 1642 by Phuntso Namgyal (Namgyal dynasty), and merged lately into Indian union in 1975. The kingdom was divided into 12 administrative regions known as Dzongs (Sinha, 2005). At present, of the three major communities – Lepchas, Bhutias and Nepalis/Nepalese – the first was believed to be earliest settlers of Sikkim (Gowloog, 2013; Sinha, 2005). In terms of religious beliefs, the Nepalese comprising about two-thirds of total population follow Hinduism. Lepchas continued to be staunch believers of Shamanism. Although it has got mixed up with Buddhism, their tradition is still being preserved, while the Bhutias practise Tibetan Buddhism. In the recent past, conversion to Christianity has been witnessed in rural areas, especially among Lepchas (Gowloog, 2013). Though English is used as official language, Nepali is commonly spoken in the state now. Hindi is also commonly spoken and understood. Several other dialects such as Bhutia, Lepcha, Limbu (Limboo or Subba), Sherpa, Rai, Tibetan and Tamang are also found and spoken in the state.[1]

Geographically, Sikkim lies on the northeastern part of India, sandwiched between Tibet (China) on its north and northeast, Nepal on its west and Bhutan on its east (see Figure 13.1). It covers an area of 7,096 sq. km. It once covered over a larger territory, much larger than the present area that touched Thang La in Tibet (China) in the north, Tagong La near Paro in Bhutan in the east and Titalia on the borders of West Bengal (India) and Bihar (India) in the south. The western border Timar Chorten on the Timar river in Nepal (Bhatt and Bhargava, 2005). In terms of population, it is estimated at 6.07 lakhs as per 2011 population census; it is the least populous state in the country. Nepalese constitutes around 65 per cent of the total population, followed by Bhutias and Lepchas with around eight per cent and seven per cent, respectively.

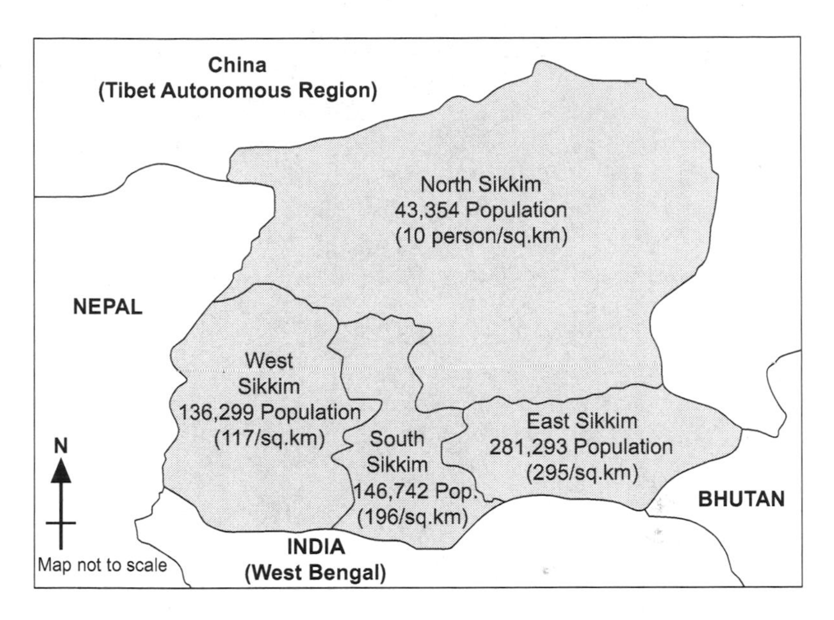

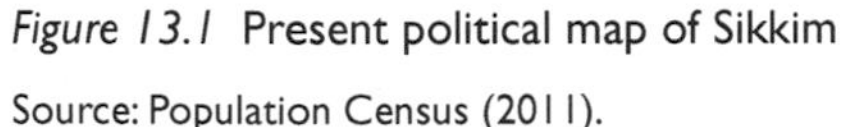

Figure 13.1 Present political map of Sikkim

Source: Population Census (2011).

The kingdom was constantly invaded by neighbouring countries such as British India, Tibet region of China, Bhutan and Nepal, from as early as seventeenth century. As a result of this, its demography has changed significantly, which led to a new sociocultural milieu. In recent past, there has been a noticeable rise in polarisation and politicisation of ethnic groups as they compete intensely to control political power and exert cultural influence. How had erstwhile dominant groups – Lepchas and Bhutias – reduced to minorities and how do they negotiate, at present, politically and economically with majority Nepali community are discussed in this chapter.

Theoretical framework

According to Kohli (1998), in any developing country, the state-society traits have significantly contributed to political conflicts because – (a) their cultural conditions do not readily mesh with imported model of democracy; (b) considerable state intervention is inherent in overall design of late development but this structural trait generate problems when democracy is introduced; (c) democratic institutions are weak and

(d) the introduction of competitive elections and mass suffrage amidst weak institutions generate more pressures towards more equal distribution of power in the society. Likewise, Ganguly (2009) underlines the following causal conditions which have combined in different ways to produce ethnic conflict in India. They are – (a) the fear that assimilation could lead to cultural dilution and the unfulfilled national aspirations; (b) the process of modernisation; (c) the unequal development, poverty, exploitation, lack of opportunity and threats to existing group privileges and (d) the political factors such as endemic bad governance, anti-secular forces, institutional decay and vote-bank politics. Furthermore, Olzak and Nagel (1986) underline the basis for ethnic mobilisation – (a) urbanisation increases contact and competition between ethnic populations; (b) expansion of industrial and service sectors increases competition for jobs; (c) development of peripheral regions or the discovery of resources in a periphery occupied by an ethnic population and (d) processes of state building (including those following colonial independence) that implement policies targeting specific ethnic population increase the likelihood of ethnic collective action.

Ethnic groups who use ethnicity to make demands in the political arena for alteration in their status, economic well-being, civil rights and educational opportunities are also engaged in interest groups' politics, and the key factor that creates ethnic consciousness is not an emotional or psychological, but a political one (Brass, 1991). Many of the conflicts are sustained through the illusion of a unique and choice-less identity, because identity is an important element that shapes nationalism and societal development processes (Sen, 2006). When ethnic identity is given the highest importance there is likely to be a basis for political mobilisation (Gurr, 2002).

Ethnicity and state formation of Sikkim

Lepchas, the earliest settler, call themselves 'Rong', the people living in ravines and follow traditional religious beliefs. They are also known as 'Meri' and 'Monpa'. The name Monpa is apparently given to them by the Bhutias – the higher altitude dwelling community, as this term refers to lowlanders. Their eastern neighbours – the Drukpas called them as Meri. The Lepchas came in close contact with Drukpas after Kalimpong (presently falls under West Bengal) was conquered by Bhutan in 1706, following a war between the two communities (Gowloog, 2013). In Tibetan language, women are called Kyeme (Kye – being; Me – lower), implies weaker/lower men in the society. Similarly, Monpa or Menpa implies people of the lower country (Men – lower; Pa – belonging). Therefore, in Tibetan language, the term Monpa has got more of status connotation rather than spatial (high or lowlanders) one. The same was applied to *Monpa* tribe of Arunachal Pradesh as well that they were treated as subordinate group by Tibetans.[2]

Bhutia/Bhotia/Bhote or Ngalops were believed to have migrated from Tibet in the fourteenth century (Pletcher, 2011), or still as early as thirteenth century (Gerke, 2012; Subba, 2011). They constitute a majority of the population of Bhutan now, where they live mainly in western and central parts of the country. Most Bhutias practice a form of Tibetan Buddhism, and more specifically are the followers of Drukpa subsect of Kagyu (Bka'-brgyud-pa), which is one of the two (of four) branches of Tibetan Budhism.[3] According to Gowloog (2013), *Drukpa* is the community which we called Bhutia of the present-day Sikkim, but the same claim was not accepted by the Sikkim Survival Council, the civil right group (SSC, 2014). As per the opinion of Gowloog (2013), *Drukpas* were originally migrated from Tibet (China), routed through Bhutan.

As of the institutionalising state politics, Lepcha leader Thokeng Tek was instrumental in the coronation of Phuntso Namgyal in 1642 by associating himself with Tibetan immigrants (Sinha, 2005). As a result, Bhutias became politically very influential (Gowloog, 2013). All the land under the kingdom that belonged to king were distributed to loyal landlords known as Kazis, and they were empowered to appoint headman who could rent land for cultivation and dwellings (Deo and Duncan, 2011).

Table 13.1 Decadal population growth of Sikkim

Decade	*Decadal growth (%)*
1901–11	48.98
1911–21	–7.05
1921–31	34.37
1931–41	10.67
1941–51	13.34
1951–61	17.76
1961–71	29.38
1971–81	50.77
1981–91	28.47
1991–01	33.07
2001–11	12.36

Source: Population Census (2011).

In terms of geopolitics, from the early seventeenth century, following frequent invasions mainly from Nepal and Bhutan, considerable portion of the kingdom was annexed – the lower Teesta basin by the Nepalese, and Bhutan occupied the eastern parts of Teesta, including the present-day Kalimpong and Jelepla ranges along Sikkim's northern and eastern boundaries (Bali, 2003). In 1816, the kingdom regained its territory following an intervention by the British, in return of their support during the Anglo-Nepalese war (1814–16) between the kingdom of Nepal and the British India. Subsequently, it became a *de facto* protectorate of the former. In 1835, the British gained control of Darjeeling (the then part of Sikkim), and as a result, within no time, developers, shopkeepers, tea planters and hordes of labourers began to pour in this newly established hill station in search of opportunities (Sinha, 2005). Subsequently, a land lease system was introduced thereby encouraging Nepali immigration, and the forestlands were allocated to them for agricultural development so that revenue could be enlarged (ibid.: 278).

In the beginning of nineteenth century, a new phase in the history of kingdom was introduced with the Anglo-Gorkha War and thereby the signing of the Treaty of Sugauli in 1815 between Nepal and East India Company, followed by the Treaty of Titaliya in 1816 between Sikkim and East India Company. Thus, the year 1816 was landmarked in the relationship between British and Sikkim, otherwise the British influenced in the kingdom. When the relationship between the king and the European officers deteriorated, a military expedition was sent from Darjeeling on 1st February 1861, reached the capital *Tumlong* (erstwhile capital of Sikkim) in early March and imposed an agreement (the Tumlong Treaty). It guaranteed opening of trade between British territories and Sikkim and removal of all restrictions. The new king of the kingdom also offered possible help to the British authorities in their efforts to develop trade with Tibet. No armed forces belonging to any other country shall pass through the kingdom 'without the sanction of the British Government'. In short, Sikkim had virtually been turned into a princely native state and a British Residency was set up in the new capital Gangtok. A political officer, appointed in 1889, was assisted by a few advisors, and hence the king virtually became a nominal head of the kingdom. A British military expedition was sent to *Lhasa* in 1904 in which the Tibetans were forced upon to recognise British over-lordship of Sikkim and to open trade relations with India (Goldstein, 1989). The power and role of the king were considerably reduced merely into the administration through this political change (LSI, 2009).[4]

On the other hand, within the ethnic communities, following a long hegemony and the denial of equal political representation to its majority Nepali population by the Bhutia monarchy, the Sikkimese political parties (mainly Nepali) started a movement for curbing the legitimacy of the monarchy (Das, 2014). The process led to unrest in the kingdom. The Nepali onslaught on Sikkim led to a bad blood between the Bhutias and the Nepalese, and there was another reason for this mutual distrust – while Bhutias looked to Tibet as their political, religious and cultural fronts, Nepalese were of Hindu orientations in such matters and tended towards Nepal (Sinha, 2005). The disproportionate growth of population caused internal conflict. At the same time, following an apprehension of Chinese intervention in the kingdom, India after its independence was instigating directly or indirectly in aggravating local unrest, interfering in local politics and compelled the kingdom to depict inevitability to withstand (Das, 2014). Internal discord and the inability of the ruler to maintain control made Sikkim vulnerable (Hart, 2001). As a result of this, despite the king's resistance and fighting for self-determination, the kingdom was united with India in 1975 (Inoue, 2005; Rai, 2013). The fact is known by other neighbouring nations as well. In the opinion of Hart (2001), 'The Nepali speaking immigrants to Sikkim are generally considered by the Bhutanese authorities as responsible for the loss of that country's independence'.

From Table 13.1, one can infer that the growth of population, primarily due to immigration, had been very significant from the 1930s till 1980s during the British India and the post-independence period. Similarly, at the time of its merger, the Hindi-speaking population has also increased significantly (see Table 13.2). This proportionate increase of immigrants from within the country (from India) was higher than that of immigrants from abroad (Nepal, Bhutan and Tibet) to Sikkim in the merger period (Lama, 2001). Ethnic crisis in Bhutan[5] in the 1990s between the Nepalese and the Dzongkha (Bhutanese) led to mass exodus of Nepalese who sought refuge in India including Sikkim and largely in Nepal. This might have probably added to some extent to the decadal growth of Sikkim's population that led to more than 33 per cent in the 1990s.

Since 20th March 1950, the British paramount over Sikkim was abolished and the new dominion of India assumed responsibility over external affairs, communications and defence. In 1953, a State Council was constituted, under which seats were equally divided among majority Nepalis and minority Bhutia-Lepchas with six seats each, respectively (LSI, 2009). Further, a tripartite agreement between India, Chogyal and representatives of three political parties, namely – Janata Congress, National Congress and National Party, at Gangtok, was signed on 8th May 1973 to establish a 'fully responsible' government with a more democratic and greater legislative and executive powers for the elected representatives; a system of elections based on adult suffrage on the basis of the principle of one man one vote and the strengthening of Indo-Sikkim cooperation and interrelationship. In this process, finally, the kingdom which had defended its independence for over three centuries against powerful neighbours was united with India in 1975 and became the 22nd state of the Indian Union (Sikkim Observer, 2014).

As of the Sikkimese identity in totality, during the days of kingdom, a population register was maintained and the inclusion in the register was, like many other established countries, not restricted only to by virtue of birth. Anyone who migrated from the adjoining countries like India, Nepal, Bhutan and China (Tibet) were granted the privilege of citizenship option, of course based on some minimal tenure of continuous stay in the kingdom. Once the name was included in the register, a citizenship certificate called 'Sikkim Subject Certificate' was issued. When the kingdom united with India, the register was handed over to India and it became basis for beneficiary list to provide certain special privileges and protections, for example income tax exemption, reservation in the government jobs, right for trade licenses and so on. Further, the descendants of these citizens (Sikkim Subject Certificate holders) were given equal or similar privileges and issued a certificate called 'Certificate of Identification' by the state government. The certificate holders are in true sense termed as the Sikkimese. This was a holistic approach that had been thought of and adopted by the then visionary leaders during the merger period, to avoid further complication of immigrants in Sikkim. Nevertheless, the policy is not free from criticism. It has also become a victim of contamination with dubious members. According to an estimate, there are 150,469 fake citizenship card holders (31,180 Sikkim Subject Certificate and 119,289 Certificate of Identification card holders).[6]

Ethnic groups and party politics

As also mentioned earlier, Sikkim is inhabited by different ethnic groups, in which Lepchas, Bhutias and Nepalese are the major ones. The Lepchas are considered to be the oldest inhabitants, believed to have originated from the foot of Mount Kanchanzungha (Gowloog, 2013), from the hills of Assam as Sinha (2005) opined, while the Bhutias emigrated from Tibet (China) in the thirteenth century (Gerke, 2012; Subba, 2011) and fourteenth century (Pletcher, 2011). The Nepalis are the most numerous group now, migrated from Nepal, which began in eighteenth century, intensified during colonial rule (Subba, 2011).

The Lepchas have not only slowly turned into an insignificant minority but also gradually lost their language, land, costumes, food habits, and even rites and rituals. The irony of the fact is that any attempt to bring them together under a common platform for protection of their culture and traditions was affected by a strong sense of divide within them between the so-called Buddhist Lepchas, who began to see themselves as true bearers of Lepcha culture and tradition, and the Christian Lepchas, who considered themselves superior to their Buddhist counterparts (Gowloog, 2013).

As population census 2011 by linguistic groups in Sikkim is not available in the public domain, the 2001 census is depicted in Table 13.2: the Nepali speakers comprised 62.6 per cent, and Bhutia speakers came in the second position with 7.7 per cent. In the third position, Hindi-speaking communities registered with 6.7 per cent and the Lepcha speakers have gone down to fourth position with 6.6 per cent. According to Turin (2008), the Nepali speakers comprised 67 per cent, followed by Bhutias with 10 per cent and Hindi speakers and Lepcha consisted of seven per cent and six per cent, respectively. Though there has been some

Table 13.2 Population of different linguistic groups in Sikkim (as per 2001 census)

Linguistic group	*Population*	%
Nepali	*338,606*	62.6
Bhutia	41,825	7.7
Hindi	36,072	6.7
Lepcha	35,728	6.6
Limbu	34,292	6.3
Sherpa	13,922	2.6
Tamang	10,089	1.9
Rai	8,856	1.6
Bengali	6,320	1.2
Urdu	2,930	0.5
Tibetan	1,977	0.4
Punjabi	1,364	0.3
Malayam	1,021	0.2
All population	**540,851**	**100.0**

Source: Mukherjee (2009).

difference in the statistics between the two sources, one can easily find a reverting trend of population equations, that an erstwhile majority Lepchas and uninterrupted Bhutias dynasty have reduced to minorities.

Caste wise, the Scheduled Tribe covers the communities of (a) Bhutia (including Chumbipa, Dopthap, Dukpa, Kagatey, Sherpa, Tibetan, Tromop, Yolmo), (b) Lepcha, (c) Limboo/Subba and (d) Tamang having a total population of 206,360, which is 33.8 per cent of the state's total population. The Bhutia comprised 63.1 per cent of the total Scheduled Tribes population followed by Lepchas with about 36.4 per cent in Sikkim (2001 census).

After Sikkim merged with India, the Nepalese gained control over the political system in the state. For instance, Tibetan, the official language till 1975 (Subba, 2011), was replaced by Nepali, Bhutia and Lepcha languages through the Sikkim Official Language Act 1977 (Subba, 2008). The first ever detailed study on linguistic survey of Sikkim – Languages in Education mentioned:

> There are effectively three lingua-franca in Sikkim, they are – Nepali, English and Hindi and all of which operate in different functional domains of use, yet constantly interact with one another. The pragmatic utility of all three languages are – Nepali in the bazaar, English in school, Hindi on television and in Central government offices – prevents any one of them from becoming overly dominant. Sikkim can experience an explosion of plurilingualism. In Sikkim, ethnic and linguistic identities are not oppositional, rather they are more incorporative.
>
> (Turin, 2008)

As of the population by religion (see Table 13.3), Hindu occupies the largest share with 61 per cent of total population, constitutes primarily of Nepalis and some sections of recent immigrants from India. Though Buddhism being the religion of the erstwhile rulers, its share has gone down; it comes in the second position with 28 per cent. It composes mainly of Bhutias, Lepcha (of course, Lepcha follows a mixture of Buddhism and Shamanism) and negligible sections of Nepalis. As of the Christianity, majority of them are from Lepcha community, and in the recent past, negligible numbers of Nepali, Bhutia, Tamang and Limbu/Subba have also converted into it. In fact, the Namgyal dynasty did not permit Christian missionaries to operate till 1975; it is rather a recent phenomenon (Gowloog, 2013). These components of population and religion are the primary basis for politics and ethnic mobilisation at present.

Table 13.3 Population by major religions in Sikkim (as per 2001 census)

Religion	*Population*	*% of total*
Hindu	329,548	60.9
Buddhist	152,042	28.1
Christian	36,115	6.7
Others	12,926	2.4
Muslim	7,693	1.4
Religion not stated	1,168	0.2
Total population	**540,851**	**100.0**

Source: India Census (2001).

Party politics started in the 1940s coincided with India's freedom movement. On the one hand, the Sikkim State Congress (SSC) supported by the Indian National Congress (INC) was set up in 1947 with the objectives of abolition of monarchy, establishment of an elected government and accession of Sikkim to India. An attempt was made to change the election system from confessional system to one man one vote principle, and its main constituents were the Nepali language speakers (Rai, 2013). On the other hand, the Sikkim National Party (SNP) was set up in 1948 with the backing of the Chogyal and his supporters in order to protect the monarchy and against the accession of Sikkim into India (Pletcher, 2011). According to Bareh (2001), the movement of SNP was against the accession on the three main following grounds – (a) Sikkim has closer affinities with Bhutan and Tibet historically and culturally; (b) Sikkim was not a part of India according to its geography and ethnicity and (c) being a Lamaist kingdom, Sikkim is distinct from India.

Under pressure from SSC, the Chogyal conceded to abolish the lessee system (land settlement), disbanded unpaid labour and judicial powers of the Kazis in 1948 (Sinha, 2005) and hence, they were stripped- off their economic and political powers. In the election of the first State Council held in 1953, the SSC won all the six Nepali seats, while the SNP won all the six Bhutia–Lepcha (BL) seats. A split in SSC led to the formation of the Sikkim Swatantra Dal (SSD) ahead of the election of 1958. In that election, the SSC won six Nepali seats as well one BL seat, while the SNP won five BL seats and one Sangha (the associations of the monks) seat. Further, the dissidents of SSC set up the Sikkim National Congress (SNC) in 1960 which became the single largest political party winning eight seats out of the 18 seats in the election of 1967. In the election held in 1970, the SNP became the single largest political party by winning nine seats.

In 1972, the SNC and the Janata Party (JP) came together to set up the Sikkim Janata Congress (SJC). With the influence of India, the first Sikkim assembly was formed through the election held in 1974 with 32 members. SJC won 31 seats out of 32 in the elections held in 1974, and finally united with the INC. In reality, the legislative assembly election of 1974 was held under the framework of Indian constitution. As of the seat sharing among the communities in Sikkim, of the total 32 seats, 15 seats were reserved for Nepalese of Sikkimese origin, another 15 for the BL and one seat each for the Sangha and Scheduled Castes. The SSC led by Kazi Lhendup Dorjee won 31 seats and one seat went in favour of SNP, and eventually Dorjee became the chief minister.

The Sikkim Janata Parishad (SJP) founded by Nar Bahadur Bhandari in 1977 won assembly election held in 1979 and Bhandari became the chief minister. The election was held in accordance with the new parity formula, that the Representation of the People (Amendment) Act of 1979 abolished the reserved seats for Nepali of Sikkimese origin and provided seats in the legislative assembly as – 12 reserved seats for Sikkimese of BL (BL seat or Schedule Tribes reserved) origin,[7] two seats for Scheduled Castes,[8] one seat for Sangha and the rest 17 seats were unreserved.[9] Interestingly, the seat set aside for the Sangha has always been filled by the BL community.

In May 1984, the Bhandari-led ministry was sacked allegedly on grounds of corruption charges which led to imposition of the central rule (president' rule). B.B. Gurung of the Congress party was appointed as the chief minister, but it lasted just two weeks (11–24 May) leading to the reimposition of the central rule. Meanwhile, Bhandari dissolved the SJP to set up the Sikkim Sangram Parishad (SSP), successfully fought and won parliamentary election held in December 1984. The party won 31 seats in the assembly election held in 1985. Bhandari vacated Lok Sabha seat to become the chief minister. His wife Dil Kumari Bhandari was elected unopposed for the lone parliamentary (Lok Sabha) seat vacated by him. His party once gain swept the elections held in 1989 winning all the seats. But, intense factional politics brought down the ministry in 1994 and Sanchaman Limboo became the chief minister for the remaining term (lasted for 179 days).

In 1993, Pawan Kumar Chamling, a former influential leader of the SSP, set up the Sikkim Democratic Front (SDF). In the election held in 1994, the new party won altogether 19 seats, 10 seats by the SSP and three seats by the Congress party. A new ministry under Chamling was constituted in December 1994. In September 1996, six members of the

SSP defected to the SDF. In the election held in 1999, the party won 25 seats and seven seats by the SSP. The party again became stronger winning 31 seats in the election held in 2004. In the election held in 2009, the party won all the seats and 22 in 2014. About a decade later, Prem Singh Tamang, also known as PS Golay, who was a close associate of Chamling, founded the Sikkim Krantikari Morcha (SKM) in 2013. It soon offered strong opposition to the ruling SDF. In the legislative assembly election held in 2014 it won 10 seats, constituting about 40.8 per cent votes. The election showed a straight fight between the two.

Several political parties including major pan-India and local parties (including registered/unregistered ones) have been contesting assembly and parliamentary elections since 1979. In the legislative assembly election held in 1979, national parties like Congress party secured just 15.55 per cent of total votes, as compared to 67.94 per cent votes secured by local parties; Independent candidates got 16.50 per cent. In 1985, national parties secured 25.46 per cent of total votes polled, while the local parties got 62.65 per cent. The Independent got 11.88 per cent votes. In 1989, national parties secured 18.05 per cent votes, while local parties got 79.21 per cent votes (and the Independent secured 2.7 per cent). In 1994, national parties secured 15.33 per cent of the total votes polled, while local parties secured 78.76 per cent (Independent 5.91 per cent). In 1999, national parties secured just 3.86 per cent, while local parties won 94.20 per cent (and the Independent got 1.94 per cent). In 2004, national parties polled 26.54 per cent of valid votes, while local parties got 71.71 per cent (and the Independent secured 1.76 per cent). In 2009, national parties secured 28.95 per cent of total votes polled, while local parties got 69.45 per cent (Independent got 1.60 per cent). In totality, local parties have been comfortably winning elections for a long time in which between 1979 and 2009, the average vote share was about 74 per cent as compared to 17 per cent secured by national parties. The share of votes polled by national political parties was lowest in 1999 (3.86 per cent). The votes polled by Independent candidates have significantly reduced since 1989. Unlike in other states, political party in power at the centre (Delhi) could not influence Sikkimese electoral politics. At the same time, except for the ruling party, most political parties were not firmly rooted and were active only during election period.

Issues and contenders

In a real sense of the term, the insider and outsider issue arises only when the immigrants cause in large-scale deprivation, displacement and discrimination of the native communities (Srikanth, 2014). Disproportionate increase in population of one or two communities in a multicultural society often leads to contestation between them. Strident identity assertions and articulation of grievances in terms of the 'others' have given rise to contestations among the communities sharing the same habitat and yet defined separately by their distinct sociological and anthropological markers. How do these issues are handled in Sikkim is the crux of this chapter.

Thirty-two constituencies constitute its legislative assembly, of which 17 constituencies are unreserved, while the remaining 15 are reserved (12 for Bhutia–Lepcha origin, two for Scheduled Castes and one for Sangha) under the provisions of the Representation of the People (Amendment) Act of 1979. It states that:

(a) in the case of a seat reserved for Sikkimese of Bhutia–Lepcha origin, he/she is a person either of Bhutia or Lepcha origin and is an elector for any assembly constituency in the state other than the constituency reserved for the Sanghas;
(b) in the case of a seat reserved for the Scheduled Castes, he/she is a member of any of those castes in the state of Sikkim and is an elector for any assembly constituency in the state;
(c) in the case of a seat reserved for Sanghas, he is an elector of the Sangha constituency; and
(d) in the case of any other seat, he/she is an elector for any assembly constituency in the state.

If the constituencies are allocated in proportion to the size of population, the Nepalese will secure at least two-thirds of the seats in the legislative assembly. Even in reserved constituencies (except three in North Sikkim), the Bhutia–Lepcha voters do not necessarily constitute a majority of voters. Under current system, the Nepali legislators constitute the single largest group in the legislative assembly enjoying unified control of the legislature and hence the bureaucracy. On the other, despite the existing reservation, the pressure groups such as the Sikkim Lepcha Bhutia Apex Committee alleged that the Lepcha and Bhutia are under-represented in the council of ministers and other important bodies. In addition, they allege that their share of employment in the public sector in total employment has declined substantially. However, they did not give specific details of the alleged claim.

Although Limboos (Subbas) and Tamangs are also recognised as the Scheduled Tribes, the reservation of seats in the legislative assembly is granted on community basis and not as per their tribal status, as in other parts of the country. As a result, they are not eligible to contest from reserved seats. It is therefore natural for them to demand suitable reservation in the legislative assembly.

In view of this, the state government has proposed to increase the number of seats in the legislative assembly from the existing 32–40. As a consequence, the Bhutia–Lepcha groups sought to increase their representation from 12 to 16. On the contrary, the Nepalese wanted allocation of seats in proportion to total population of the state. If so, they will constitute at least two-thirds of the legislative assembly,[10] simply because they are the most numerous group. The Nepali pressure groups such as the Gorkha Apex Committee have maintained that the existing reservation of seats is 'unfair' and 'unconstitutional'. It wanted the allocation of seats on the basis of population ratio. In this context, since Limboos and Tamangs also deserve to get reservation in the legislative assembly as in other parts of the country where seats are reserved for Scheduled Castes and Scheduled Tribes in the legislative assemblies. Another contentious issue is the Article 371F of the Constitution of India which stated that:

> all laws in force immediately before the appointed day in the territories comprised in the State of Sikkim or any part thereof shall continue to be in force therein until amended or repealed by a competent legislature or other competent authority . . . for the purpose of facilitating the application of any such law as is referred to in clause (K) in relation to the administration of the State of Sikkim and for the purpose of bringing the provisions of any such law into accord with the provisions of this Constitution, the President may, within two years from the appointed day, by order, make such adaptations and modifications of the law, whether by way of repeal or amendment, as may be necessary or expedient, and thereupon, every such law shall have effect subject to the adaptations and modifications so made, and any such adaptation or modification shall not be questioned in any court of law.

The Bhutia–Lepcha groups who feel protected by Article 371F want to retain old laws since they consider introduction of new laws will outlive the relevance of old laws. For instance, they had opposed the inclusion of Dukpa/Drukpa, Kagatey, Sherpa, Tibetan and Yolmo into Bhutia ethnic group under the provisions of the Constitution (Sikkim) Scheduled Tribe Order of 1978 since they (Bhutia–Lepcha) considered them (communities mentioned earlier) as 'Nepali colonists' of people of Nepali origin (SSC, 2014). Furthermore, they also wanted the agreement that united Sikkim with India be recognised as the basis for future political arrangements including Article 371F. The Article 5 of the tripartite agreement between India, Chogyal and representatives of three political parties on 8 May 1973 states that:

> The system of the election shall be so organized as to make the assembly adequately representative of the various sections of the population . . . Care being taken to ensure that no single section of the population acquires a dominating position due mainly to its ethnic origin, and that the rights and interests of the Sikkimese of Bhutia Lepcha origin and of the Sikkimese of Nepali which includes Tsong and Scheduled Castes origin are fully protected. (Sinha, 2005: 289)

The BL groups alleged that in the name of democracy, there has been a 'departure' from the 'real spirit and objectives' of the said agreement. They wanted restoration of seat reservation for Nepalese of Sikkimese origin in the legislative assembly. Additionally, they also wanted special reservation for BL in the Panchayat and Municipal bodies.

They were apprehensive of the new legislations including the Sikkim Regulation of Transfer of Land Bill of 2005. Its main objective is 'to make provision for the regulation of transfer of lands, covering wider section of the population in the State and other matters connected therewith. Whereas the old laws on transfer of land catered to certain section of the population in the State; and whereas it has been considered expedient to have law regulating transfer of land covering wider sections of the population in the State and strengthen the existing law further.' The Act prohibits the transfer of land belonging to any person, by way of sale, gift, exchange, mortgaged or sublet with possession shall be valid in favour of a person who is not an agriculturist. The Sikkim Bhutia–Lepcha Apex Committee (SBLAC) condemned this legislation as 'murder of democracy' primarily because they wanted to enjoy due protection under the old laws, enacted between 1926 and 1973. They felt that new legislations will further dilute the relevance of old laws which they considered are protected by the Article 371F of the Constitution of India.

Further, they were also alarmed by what they called 'distorting history' referring to the alleged 'errors' relating to the Buddhist festival of Pang Lhabsol in a government-sponsored study of 2004. It alleged that the festival was never celebrated on 2nd September as mentioned in the study instead it was from 8 to 15 day of the seventh month of the Sikkimese lunar calendar corresponding to the month of late August or early September. They have also raised concern over the impact of the developmental projects on historical and sacred sites. In this context, a vigilante group BL Protection Force has been constituted. It had organised a convention in Gangtok in December 2013 to deliberate on the issues affecting the people. Although they invited the Lepcha–Bhutia members of the legislative assembly, none of them turned up. Besides, they also wanted an additional 10 per cent reservation in employment and education for the Lepchas, after they were recognised as a Scheduled Tribe in 2002, in addition to the benefits the community currently enjoys. Ahead of the April 2014 assembly election, the SBLAC offered to support the SKM primarily because they felt discriminated by the SDF. In short, the SBLAC (2003) had resolved:

(a) proportionate increase of BL seats in case of increase of the number of legislative assembly seats from 32 to 40;
(b) reservation of seats for BL at the Panchayat level on the basis of reservation principle as applied in assembly seats;
(c) reservation of seats in higher studies and appointment on the basis of reservation principle as applied in assembly seats; and
(d) delimitation of assembly constituencies on the basis of Census 2001 ensuring that BL voters remain in majority in the reserved BL constituencies.

Further, the SBLAC wanted stringent management of Sikkim's borders with Bhutan and Nepal to check illegal immigration. Tseten Tashi Bhutia, a prominent leader, has once stated:

> Presence of and incoming large number of illegal emigrants or infiltrators from neighbouring countries which have caused immense social, political, economic and ethnic imbalance in the State . . . despite the best efforts by few social organizations to bring the focus on the serious consequences of growing influx, the policy-makers have not been able to check this unabated flow . . . The entry of foreign nationals from the two neighbouring countries existed ever since Sikkim became a protectorate Kingdom of British India and has been more pronounced since Sikkim became a State of India. The growth of population has been attributed to unabated immigration of people from outside the State and neighbouring countries.[11]

Argument and concluding remarks

Different ethnic groups of Sikkim had historically migrated from different places, and have been living together in harmony. The diverse ethnicities sharing the same space are defined separately by their distinct social constructs. In reality, there are advantages for ethnic groups to realign themselves. Alternatively, rising competitive forces can cause reactive mobilisation on the part of numerically advantaged groups. Political disadvantage relates to systematic limitation of access to political office; economic disadvantages are systematic denial of economic goods and opportunities (Ginsburg and Dixon, 2011: 364).

Unlike other States of Northeast India[12] such as Manipur or Assam, Sikkim does follow a balanced and inclusive approach that takes all ethnicities together. This makes Sikkim a peaceful and prosperous place. In other states, several religious or ethnic communities sharing same space have their own exclusive political aspirations which led to struggles for independent homeland from others and often demands for more political autonomy, either within or outside the states. There are number of ethnicity-based political parties in other states, especially in the region. In the case of Sikkim, the land originally inhabited by Lepchas, founded Namgyal dynasty by the Bhutia immigrants from Tibet with the help of a Lepcha chief (Little, 2010; Sinha, 2005), is now headed by a Nepali Chief Minister Pawan Kumar Chamling. He is truly a dynamic leader. Besides, Bhutia monarchy did not force their predecessor Lepchas to convert into Buddhism. This is the reason why Lepchas can preserve their identity and religion till date, albeit negligible member of the community have converted into Christianity in the recent past. Despite some political parties that have been formed on the basis of political and economic benefits, no party was formed on the basis of religion and ethnicities in Sikkim. Content analysis of party election manifestoes tells an interesting story. After Sikkim united with India, ethnically inclined political parties became irrelevant. Election manifestoes of SDF highlighted forward-looking issues such as higher per capita income, high literate, tourism destination in South and South-East Asia, control pollution and disease, best performing state of the country and efficient mountain economy. In 2014, it gave special focus on youth-oriented and rural development programmes.

The assembly seat arrangement[13] clearly reflects inclusive approach that 41 per cent (including Sangha seat) of the total assembly seat is reserved for 16–20 per cent of BL population. Sikkim is the first state in the region to reserve assembly seat for Buddhist monk (Sangha) community. The well-thought policy, considering its multicultural society could take all the communities together. Therefore, Nepalese demand for reallocation of assembly seats in proportion to the total population of the state, in reality, is not an illogical, rather a legitimate one. When it comes to a unified ethnicities of Bhutias and Lepchas, their primarily contention is to counteract the numerically dominant Nepali-speaking people. Despite criticism from different quarters for different reasons, the SDF, ruling for five consecutive terms since 1994, have adopted accommodative approaches to make Sikkim one of the most peaceful and prosperous states in the country.

Although the minorities are united together to pursue political objectives against the majority, their aim is not to establish domination over the majority group, only to secure special concessions from the government, not to impose religious hegemony. At the same time, the majority group also seeks to retain its hold on political and economic powers and cultural influence. Nevertheless, the differences between ethnicities are being balanced by the spirit of accommodative, sharing and humanity approaches of the communities living in the state. Reservation of seats is one way to ensure representations of minorities in the legislative assembly. As discontentment arises, on the issues of reservation, the government should seriously strive to increase the constituencies of the legislative assembly and recognise the special needs of the minorities.

What is unique to Sikkim is that despite polarisation of ethnicities, a strong Sikkimese identity is prevalent which clearly overrides ethnic identity differences. Ethnically inclined political parties such as SSC and SNP have been completely sidelined by the inclusive model of the political parties such as SDF. This is an extraordinary achievement and exceptional, unlike the ethnically inclined political parties in the other parts of Northeast India. In this context, political parties and their leaders play a major role. Effective leaders are important to solve problem of how to organise collective effort; consequently, they are the key to organisational effectiveness, in the context of the state of Sikkim.

Notes

1 Retrieved from http://www.bharatonline.com/sikkim/culture/languages.html.
2 This section was contributed by Mr Nyima Tenzin, research scholar, Department of Economics, Sikkim University. Being a Tibetan origin, born and brought up in Arunachal Pradesh and settled in Sikkim, he is well aware of the facts and genesis of the issue.
3 Excerpted from http://www.britannica.com/EBchecked/topic/64255/Bhutia.

4 This section is excerpted from the Linguistic Survey of India (2009): Section I – Brief History of Sikkim.
5 In September 1990, against the Bhutan's policy of Bhutanisation, thousands of ethnic Nepalis, especially the southern part of the country (bordering Sikkim, West Bengal and Assam), protested in large scale. The government's response to the demonstrations was reportedly swift and harsh, and the months that followed saw widespread arbitrary arrests, ill treatment and torture, followed by an exodus from the country of thousands of ethnic Nepalese from southern Bhutan.
6 An open letter sent to chief minister of Sikkim on fake 'Sikkim subjects' issue', by Bhutia, Tseten Tashi (Ex-MLA of Sikkim) on 25/08/2014. Retrieved from http://jigmenkazisikkim.blogspot.in/2014_08_01_archive.html.
7 The constituencies reserved for BL are Yoksam-Tashiding, Rinchenpong, Daramdin in West district, Barfung, Tumen-Lingi in South district, Gnathang-Machong, Shyari, Martam-Rumtek and Gangtok in East district, and Kabi Lungchuk, Djongu and Lachen Mangan in North district.
8 The four Scheduled Castes – Kami, Damai, Sarki and Majhi – belong to the Nepali community.
9 In 1979, fifteen seats reserved for Nepali community were abolished, it was started in 1953. The fifteen seats reserved for BL have been reduced to twelve, retaining one seat for Sangha (Buddhists monk group) and two seats for Scheduled Castes.
10 Refer to National Commission for Scheduled Tribes (http://ncst.nic.in/writereaddata/linkimages/agenda171109-V6813613364.pdf).
11 Refer to the document of *Sikkim Bhutia-Lepcha Apex Committee*, 20th December 2003. Retrieved from http://www.siblac.org/chronicle_2003.html.
12 India's NER consists of seven states – Assam, Arunachal Pradesh, Manipur, Meghalaya, Mizoram, Nagaland and Tripura. Later on, the state of Sikkim joined in the region's fabric in 2002 and with it, now NER consists of eight states.
13 Presently, in Sikkim legislative assembly, the seat arrangement is made as: twelve reserved seats for BL, two seats for Scheduled Castes, one seat for Sangha and the rest of the seventeen seats were declared as general seats.

References

Bali, Y. (2003). *Pawan Chamling – Daring to be different*, Information and Public Relations Department, Gangtok (India): Government of Sikkim.

Bareh, H. (2001). *Encyclopaedia of North-East India: Sikkim*, New Delhi: Mittal Publications.

Bhatt, S.C. and Bhargava, G.K. (eds) (2005). *Land and People of Indian States and Union Territories – Volume 24 Sikkim*, New Delhi: Kalpaz Publication, p. 19.

Brass, P.R. (1991). *Ethnicity and Nationalism: Theory and Competition*, New Delhi: Sage Publications.

Das, S. (2014). 'Sikkim the Place and Sikkim the Documentary: Reading Political History through the Life and After-Life of a Visual Representation', *Himalaya*, 33(1): 40–56.

Deo, N. and Duncan, M. (2011). *The Politics of Collective Advocacy in India: Tools and Traps*, Bloomfield (USA): Kumarian Press.

Ganguly, R. (2009). 'Democracy and Ethnic Conflict', In Ganguly, S., Diamond, L. and Plattner, M.F. (eds), *The State of India's Democracy* (pp. 45–66), New Delhi: Oxford University Press.

Gerke, B. (2012). *Long Lives and Untimely Deaths – Life-Span Concepts and Longevity Practices among Tibetans in the Darjeeling Hills, India*, Leiden, The Netherlands: Koninklijke Brill.

Ginsburg, T. and Dixon, R. (eds) (2011). *Comparative Constitutional Law*, Cheltenham: Edward Elgar Publishing Limited.

Gowloog, R.R. (2013). 'Identity Formation among the Lepchas of West Bengal and Sikkim', *Studies of Tribes & Tribals*, 11(1): 19–23.

Gurr, T.R. (2002). *Peoples versus States: Minorities at Risk in the New Century*, Washington, DC: United States Institute of Peace Press.

Hart, J. (2001). 'Bhutan: Conflict, Displacement and Children', Discussion Paper, Refugee Studies Centre, Oxford: Oxford University.

Hogg, M. and Abrams, D. (1988). *Social Identifications: A Social Psychology of Intergroup Relations and Group Processes*, London: Routledge.

India Census (2001). *India Census 2001– State-Wise Religious Demography*. Retrieved from http://www.crusadewatch.org/index.php?option=com_content&task=view&id=580&Itemid=27.

Inoue, K. (2005). 'Integration of the North-East: The State Formation Process', In Murayama, M., Inoue, K. and Hazarika, S. (eds), *Sub-Regional Relations in the Eastern South Asia – With Special Focus on India's North Eastern Region*. Chiba, Japan: The Institute of Development Economics, pp. 16–31.

Kohli, A. (1998). 'Can Democracies Accommodate Ethnic Nationalism? The Rise and Decline of Self-Determination Movements in India', In Basu, A. and Kohli, A. (eds), *Community Conflicts and the State in India*, Delhi: Oxford University Press, pp. 7–32.

Lama, M.P. (2001). *Sikkim Human Development Report 2001*, New Delhi: Social Science Press.

Little, K. (2010). 'From the Villages to the Cities – The Battlegrounds for Lepcha Protests', *Transforming Cultures, e-Journal*, 5(1): 84–111.

LSI (2009). *Linguistic Survey of India-Sikkim Part I*, Language Division of the Office of the Registrar General, New Delhi: Ministry of Home Affairs of Government of India.

Mukherjee, K. (2009). 'LSI-Sikkim, Part-I'. Retrieved from http://www.censusindia.gov.in/2011-documents/lsi/LSI_Sikkim_Part%20-II/Chapter_I.pdf.

Olzak, S. and Nagel, J. (1986). *Competitive Ethnic Relations*, Orlando: Academic Press.

Pletcher, K. (ed.) (2011). *The Geography of India: Sacred and Historic Places*, New York: Britannica Educational Publishing, pp. 247–53.

Population Census (2011). 'Provisional Population Totals Paper 1 of 2011–Sikkim', Office of the Registrar General and Census Commissioner, Ministry of Home Affairs, Government of India.

Rai, D. (2013). 'Monarchy and Democracy in Sikkim and the Contribution of Kazi Lhendup Dorjee Khangsherpa', *International Journal of Scientific and Research Publications*, 3(9): 1–13.

Sen, A. (2006). *Identity and Violence: The Illusion of Destiny*, London: Allen Lane.

Sinha, A. C. (2005). 'Sikkim', In Murayama, M., Inoue, K. and Hazarika, S. (eds), *Sub-Regional Relations in the Eastern South Asia: With Special Focus on India's North Eastern Region* (p. 275–97), Chiba (Japan): Institute of Development Economies-Japan External Trade Organization.

Sikkim Observer (2014). 'Chogyal was Placed under House Arrest Before Sikkim's Annexation', Saturday, 22–28 February 2014. Retrieved from http://jigmenkazisikkim.blogspot.in/2014/02/sikkim-observer-page-1–22–28–2014-vol-2.html.

Srikanth, H. (2014). 'Who in North-East India Are Indigenous?' *Economic and Political Weekly*, 49(20): 41–6.

SSC (2014). *Survival Sikkim*, Tadong, Gangtok: Sikkim Survival Council.

Subba, J. R. (2008). *The Evolution of Man and Modern Society in Mountainous Sikkim*, New Delhi: Gyan Publishing House.

——— (2011). *History, Culture and Customs of Sikkim*, New Delhi: Gyan Publishing House.

Turin, M. (2008). *Results from the Linguistic Survey of Sikkim – Mother Tongues in Education*, Gangtok, Sikkim: Namgyal Institute of Tibetology.

Chapter 14

Ethnic assertion in Manipur

Reflection on electoral integrity and governance

L. Muhindro Singh

Introduction

In this modern developed world, the people in Manipur[1] have been living in a very insecure and uncertain life, mainly the youths who were born after 1980. The state of affairs is very unpredictable; the psycho-fear is associated in every mind of the people in the state. The fake 'encounter' levelled by security forces is a very common phenomenon in the state. Sometimes, stage-managed encounter is also carried out in broad daylight to conceal their atrocities or wrongdoings. The armed forces carried out arbitrary arrest, torture and extrajudicial executions with impunity. The central government fails to respect human rights in Northeast India, which is well proved by the copious incidents happened in the region. Besides different acts of state, a number of non-state actors have also been threatening human rights in Manipur. The United Naga Council (UNC) made clear that they would like to be part of Greater Nagaland as demanded by the National Socialist Council of Nagalim-Isak and Muivah (NSCN-IM)[2] and strongly get involved in the electoral politics since NSCN-IM entered in political dialogue with Government of India (GoI) in 1997. It is understood that the democracy largely depends on the level of electoral system and its participation (Lijphart, 1999). Those scholars who dealt with the democracy felt to appreciate the quality of new democracies around the world, and considered the changing pattern of mass participation (Michael *et al.*, 2010). If the result of the elections cannot reflect accurately the will of the electorates, the process and the outcome will demean democracy and eventually develop flawed democracy. With these understandings, the present study tries to explore the scenario of electoral integrity and governance in the conflict zone of Manipur where ethnicity-based insurgent groups often interfere in the democratic process, which eventually leads to dysfunctional governance. Nevertheless, the specific objectives of this chapter are as follows:

- to reveal how the different ethnic groups attempt to assert their conflict situation;
- to examine the modus operandi of the ethnicity-based armed opposition groups against electoral integrity;
- to understand the existing phenomena of dual governance in the state; and
- to give an insight on how the people assume the prevailing trend of electoral politics in this conflict state.

Methods and limitations

No single method can fulfil the expected result of social science research which is agreeable to the public. For more accuracy, quantitative, qualitative and empirical methods are the primary tools that are adopted here in this chapter. The data were mainly collected from the structured questionnaire surveyed in the entire Manipur as well as other secondary sources. But many electorates and stakeholders are still hiding their faults; instead it is seemingly developing a new culture of electoral trend. In the context of electoral integrity, it is attempted to analyse the specific population size, altogether 140 samples are collected from all the nine districts of Manipur – Chandel 13; Senapati 12; Thoubal 15; Imphal East 18; Imphal West 31; Churachandpur 12; Ukhrul 11; Bishnupur 10 and Tamenglong 18. Logically, intimidation and threat might have certainly impacted on the freedom of choice, but it is very doubtful as the findings are relatively different from the assumption and witnesses. The matter of dual governance is not much emphasised, but it is just based on empirical and available news reports that have made relevance with electoral integrity.

Concept of electoral integrity

The concept of electoral integrity is understood as shared international principles, values, and standards of democratic elections which apply universally to all countries and which should be reflected at all stages during the *electoral cycle*, including the pre-electoral period, the campaign, polling day and its aftermath. Violations of electoral integrity, by contrast, constitute electoral malpractices (Pippa, 2012). For conducting free and fair elections, the concerned authority provides dos and don'ts in their respective representative of the people and model code of conducts, but many political parties, candidates and connivances have often committed illegal acts on the eve of elections; all these are malpractices. For instance, different party leaders overtly and covertly made attempt to woo/lure people in the name of collective goods or for a particular backward community. The notion of 'electoral integrity' is gaining in popular usage as an all-encompassing way to conceptualise many related problems.[3]

International principles and standards

There is no single best electoral system and as the usage of majoritarian, mixed and proportional systems serve different goals and functions, such as the priority which should be given to governance effectiveness versus inclusiveness (Pippa, 2004). Societies vary in their legal regulations and procedures governing elections, for example, in political finance regulations governing, the use of public funds for political parties and in the use of contribution or spending limits for campaigns (James, 2005). Contrasts in these matters are evident even amongst relatively similar long-established European democracies, such as Germany, Sweden, Britain and France (Karl-Heinz, 2009; Kevin, 2004; Magnus and Hani, 2011; Michael, 2011). It would therefore be inappropriate to adopt procedural standards or legal frameworks from any one society as the legitimate basis for evaluating practices elsewhere.

To maintain uniformity in all the countries, the Universal Declaration of Human Rights 1948 supported such electoral rights that all the electorates should have freedom to choose their representatives as well as to be candidature to participate in the governance. The will of the people shall be the basis of the authority of government; this will be done through the periodic and genuine elections which shall be by universal and equal suffrage and shall be held by secret vote or by equivalent free voting procedures. Agreement about the general standards which should govern the conduct of elections is further specified in the International Covenant on Civil and Political Rights, ICCPR (1996, fifty-seventh session). The most authoritative international standard to evaluate electoral integrity has been operationalised in detail in the practical guidelines issued to electoral observers, exemplified by the *Election Observation Handbook* (6th edition) published by the Organisation of Social and Economic Cooperation (Eric, 2004).

Electoral cycle

Elections should be seen as a sequential process or cycle, involving a long series of steps. Electoral observation should not be focused purely on the election day, or even on the short-term period of the official campaign. As the Administration and Cost of Elections (ACE) project suggests – the cycle involves all stages in the process of elections: from the design and drafting of legislation, the recruitment and training of electoral staff, electoral planning, voter registration, the registration of political parties, the nomination of parties and candidates, the electoral campaign, polling, counting, the tabulation of results, the declaration of results, the resolution of electoral disputes, reporting, auditing and archiving. After the end of one electoral process, it is desirable to begin the next: the whole process can be described as the electoral cycle.[4]

Ethnic assertion in the context of Manipur

Manipur has been experiencing 'ethnic assertion' in different ways and dominating in contemporary elections, especially in the hill areas. No one will deny that Manipur has dual governance: constitutional and non-constitutional, the latter being more powerful. In contemporary, many armed opposition groups have been entering in electoral politics, and politicians have nexus with them for mutual benefits. Such trends are seemingly very noticeable phenomena in electoral politics in the sense that both the central and state governments have also somehow managed it on the backdrops of existing conflict situation and claimed free and fair elections. They (non-state actors) often attempt to nominate their candidates who have been loyal to them; eventually, such coercive influence violates electoral rights of those aspirant candidates. And it also creates unusual social chaos where people are reluctant to express their freedom of thought. On one hand, ethnic civil organisations and frontal organisations supported by armed outfits are also actively involved in electoral politics in the state, more importantly in the hill areas with certain objectives. On the other, the dynamics of ethnic assertion on the eve of elections in the valley area is quite different; that is such ethnic influences are very low. Various factors associated with ethnic conflict sometimes affect electoral integrity. In such conflict areas '*bullying*' is one of the most common factors on the eve of election. The contemporary electoral rights are often violated on the eve of elections.

Political party on ethnic line

Obviously, ethnic political party has not yet emerged in the valley. However, though some hill-based political parties are much vocal on the ethnic line for their political gain, they are not so successful. The Kuki National Assembly (KNA)[5] was one of the earliest political parties which emerged in the hills and raised the grievances of the hill people and actively participated in the electoral politics though it was seemingly extinct in contemporary.[6] The party stands for preserving and improving the culture of the Kukis, freedom from external exploitation and unification of their geographical areas into a political unity (Roy, 1986). Another ethnic political party, the Naga National Party (NNP), emerged from a sense of disillusionment on the part of certain sections of the Nagas due to the absence of effective and accountable government inside the state for a long time and no proper attention to the genuine voices of the Nagas.[7] The party aims to promote emotional integration and unity, and build up the spirit of oneness among all the Nagas, to help in bringing a lasting and peaceful political solution to the age-old Naga political issue through dialogue and to maintain Naga identity. Significantly, in the recent years, Nagaland-based ethnic political party, NPF (Naga People's Front), was on the scene of electoral politics on the eve of Manipur State Assembly Elections 2012. Since then, many people drew attention on such new phenomena of ethnic politics in the hill area where even the other state's (Nagaland) chief minister, Neiphiu Rio, campaigned and ethnic assertion was more vocal.[8] From various quarters, it is observed that the NPF[9] is working with the support of one Naga insurgent group. It is also evident from various appearance and allegations. Nonetheless, some new strategy was also found in the hill areas. Earlier the NSCN-IM put up their candidate as 'Independent' but in State Assembly Elections 2012, it nominated twelve candidates through the NPF. The UNC called on the Nagas to support the NPF, making its debut in Manipur with the stated aim of uniting the Naga-inhabited area under one administrative roof.[10]

Emerging electoral trend and dual governance

Elections in the remote hill areas are not free and fair in Manipur. Most of the officials who are engaged in the election in the hill areas opined that, 'they cannot perform the electoral norms what the Election Commission instructed due to various factors, that question of insecure life is most important, so conditionally they have to compromise in a manageable way'.[11] In general, if one employee is bound to serve election duty in the hill area, first and foremost conception and priority was 'alternative means if possible'. Many political workers or protagonist in the particular polling area have often acted as authority and committed different coercive activities in favour of a particular candidate. Election officials and security forces seem to have no power in such circumstances. They cannot interfere to the unfair means in election. Interestingly, so far, none of the electorates made complained in regards to their electoral right. It is true that government machineries in the hill areas cannot work properly as there is dual governance of state and non-state actors. Most of the top officials posted here need to develop a cordial nexus with dominant non-state actor. Various incidents and appearance witness the nexus with deputy collector of the districts and armed opposition groups. Otherwise, they have to bear dire consequence such as killing, torture and kidnapping. Indeed, most of the electorates cannot do against the will of the armed opposition groups and its frontal organisations.

Inefficiency of governance has been witnessed since long back which is also by-product of power politics and vote-bank consideration leading too many ill-social practices in the society. Ultimately it generates social unrest and chaos. Impelled by the arrogance of its own power, the government uses force to bring down the agitation but the governmental action for restoration of normalcy produces its own ill effects; it generates only resentment. In such conflict zones, all the governmental functions likely are handicapped and dominated by the armed forces both from Indian Army and insurgency groups (state and non-state actors). Administration of state affairs is also undergoing parallel government by state machinery on the one hand and insurgent groups on the other. Indeed, there is no impression of governance in actual form, in particular, in hill districts. It is very unfortunate that people of this state are living in such insecurity in a democratic country. Certainly, people are in dilemma and often become victims of both the forces.

Large number of insurgent groups, so to say, most of the ethnic groups have different insurgent groups that are operating in the state. Valley insurgent groups like United National Liberation Front (UNLF), People's Liberation Army (PLA), Kangleipak Yaol Kanna Lup (KYKL) and others seem to control the valley areas. Many antisocials and culprit, including high-ranking bureaucrats, who were indulged in corruption and misbehaviour, have been punished by them accordingly. But, fund collection through illegal process is very high; donation from business communities, contractors, percentage collection from different governmental and non-governmental projects and even from monthly pay of employees is quite common in both hill and valley areas. In the same way, the NSCN-IM and numerous Kuki insurgent outfits like KNA and KNF have been controlling hill administration. Government machineries are acting on the drum beats of these outfits. House tax collection and goods tax collection from heavy vehicles on the National Highways (NH 39 and NH 53) is quite common. Government can do nothing in this regard. Actually, people have also followed the instruction given by them (insurgent groups) and otherwise the casualties/victims will be individualised. In the meantime, counter-revolutionary activities advocated by infiltrators or surrenderres and reactionaries have also been growing at the very fast pace.

Besides, most of the candidates have their own illicit forces and spread psycho-fear among the electorates by various activities.[12] Obviously, elections in hill and remote areas are witnessing abnormal election where many electorates are not able to enjoy electoral rights.[13] Not only electorates, the candidates too have been facing threatenings from various quarters. Moreover, some ethnic civil organisations who have loyal to armed opposition groups have been acting like a political party and are imposing dos and don'ts to the electorates and candidates. In such areas, political elites have nexus with insurgents and are using them for mutual benefits (Marwah, 2010). Perhaps, the conflict situation may compel to do so. Politics in the Northeast is not based on democratic norms as understood in the western-model democratic system. As money is the name of the game, most political leaders find it convenient to develop close associations with one or other insurgent groups. In this context, former Governor of Manipur, Ved Prakash Marwah, commented:

> During elections, political leaders seek support of the insurgents in the form of muscle power. They also depend on them for emotional appeal to the electorate on ethnic lines. In lieu of their support, they share part of the loot from public funds with them and intervene with the state police in their favour. This in turn has led to a breed of greedy political leaders whose main interest in politics is to make money and not serve the people. The entire political system in the region has grown around these unscrupulous politicians.[14]

Threat to party or candidate

Threats or intimidations to follow their (armed opposition groups) diktat to the particular candidates or party are not an exception in this state. It is witnessed by various reported news and appearances. Indian National Congress has alleged the NPF of using insurgents/armed opposition groups in their favour and resorting to all sorts of election-related crimes in the recently held Manipur state assembly elections, 2012.[15] The Kuki militant groups, which are under the Suspension of Operations (SoO), and the NSCN-IM, which is also under peace talks with the centre, were openly involved in elections. The Kuki militants openly supported the Congress party candidates, while NPF was supported by the NSCN-IM. Involvement of the insurgents in the elections threaten, intimidation and challenges on electoral rights is no doubt will remain so long.[16]

A conglomeration of several valley-based insurgent groups, CorCom (the Coordinating Committee), unsuccessfully targeted to defeat candidates belonging to the Congress party.[17] Admitting the involvement of the NSCN-IM in election matter, the former Home Minister P. Chidambaram said NSCN-IM is in talks with the Government of India. It is also true that their cadres along with some other organisations continue to indulge in kidnapping and sometimes violence.[18] The finding shows that most of the threats are coming up from armed opposition groups which means 50 per cent respondents hinted though 18 per cent kept mum due to psycho-fear. Besides, state-actors who ought to protect electoral rights are also among the perpetrators.

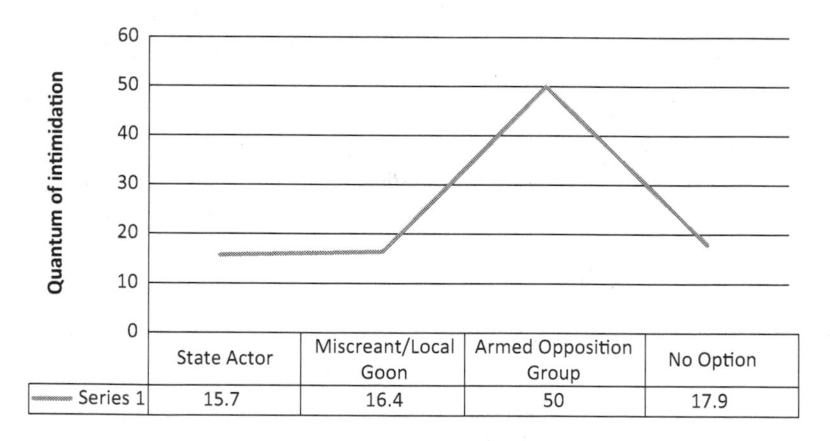

Figure 14.1 Intimidation/threat to party or candidate from any quarter

Source: Singh (2013).

Bullying to electorates

Bullying or intimidations on particular electors or electorates were also aplenty on the eve of elections. The modus operandi of intimidation might be any act of bullying or coercive force to make behavioural change of electors, for instance, intimidation to support or not to support particular candidate or party, or not to involve in any electioneering activities which is an attempt to curve freedom of electorates on the one hand. In short, it is violation of electors' right who is supposed to enjoy freedom of choice to exercise their conscience and wisdom to elect the worthy candidate. Such intimidations often appear in many parts of the state, though the frequency is accordingly different. For instance, after banning the Congress party indefinitely and issuing warnings against its workers to desist from election-related works, the CorCom launched a series of bomb attacks on Congress party offices and residence of workers.[19]

A senior functionary of the Senapati District Congress Committee was severely assaulted by NSCN-IM on 26 January 2012. Amidst the threat of NSCN-IM, a peace rally was staged at Chandel on 19 February 2012 by womenfolk demanding that they be given the freedom to freely exercise their franchise without threat and intimidations.[20] The Naga National

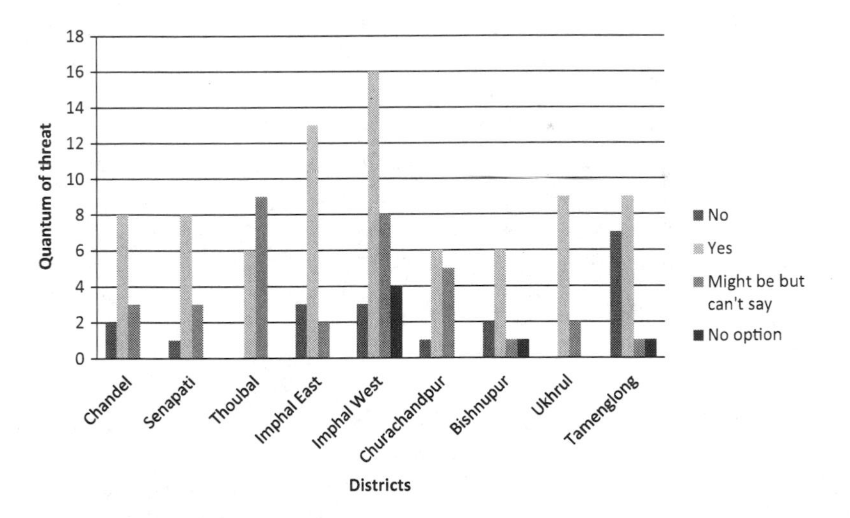

Figure 14.2 Threat to support or not to support any particular candidate/party

Source: Singh (2013).

Council (NNC) charged the Isaac Muivah leadership of shrieking their responsibility owed to the 'Naga Nation'. It alleged that the formation of the United Naga Democratic Front (UNDF) was at the behest of the NSCN-IM. It also alleged that the Nagas in Manipur are not only forced and threatened to vote for the UNDF candidate but the NSCN-IM leaders have also been abducting public leaders from Tamenglong and Ukhrul district areas for raising objections to their stances.[21]

In totality, the finding shows that unexpectedly valley area has suffered most on such bullying issues. The Imphal west district was the most affected by threat to electorates, for example, 11 per cent of the electorates faced the threat out of 58 per cent opined they have noticed such bullying on electorates while 6 per cent out of 24 per cent supported the argument that they projected as saying 'might be but cannot say'. Significantly, none of the districts are beyond these issues in the sense that the other districts except Imphal east district were also within the range of 5 to 6 out of 58 per cent who are aware of threatening on electorates. More or less bullying on electorates is cleared.

Impact of Naga movement: a review

The ethnic relationship between the Meiteis (the dominant community of Manipur) residing in the valley areas and the Nagas living in the hills has deteriorated after the June 2001 turmoil in the state following the Central Government's decision to extend the area of ceasefire with the NSCN-IM as 'without territorial limits'. The UNC in the state has made clear that they would like to be part of Greater Nagaland as demanded by the NSCN-IM. On the other hand, for the Meiteis and other ethnic groups, this became a question of defending their territory. This divide has its impact on the political set up, including a realignment in the political parties, the electioneering process and, possibly, on the outcome of the polls (Bibhu, 2010). As a new strategy, the NSCN-IM has gradually entered in the state politics directly or indirectly by condemning Delhi's delaying tactics. On the other hand, the centre has got a golden opportunity when they arrived in state politics in one sense. They had remained outside the electoral politics and banned state elections in the hill districts before the ceasefire dialogue started. Indeed, during the appearances of electoral politics in the hill areas, the Naga politicians are very seemingly to be under the control of NSCN-IM or UNC. On the eve of elections, aspirant candidates have signed an 'Undertaking' as oath to do for the Naga movement (Eight-Point Declaration).[22] The Naga issue became one of the significant as well as handy issues of every political party side by side in hill and valley. It may not be wrong to observe that politicians may not want to resolve the Naga issue and the Armed Forces (Special Powers) Act of 1958 as these are the core issue of election.

On the eve of elections, many Naga NGOs and frontal organisations, and armed opposition groups (mainly NSCN-IM) have been trying to make more united and closed cooperation by organising traditional functions and even campaigning door to door in the entire Naga-dominated hill areas to convince and support sponsored Naga candidates on the backdrop of Naga integration. Some Naga politicians and social organisations claimed that the word 'Manipur Integrity' is against their aspiration of Naga unification (unification) though some sections of Naga community do not have much interest on the integration demand. Taking the opportunity on this diabolic situation most of the politicians in hill and valley have often been made a great deal on ethnic or communal politics. NSCN-IM, on 13 July 2005, and All Naga Students' Association Manipur (ANSAM) asked all the Naga MLAs and Ministers in Manipur to resign. They declared that Naga identity and dignity cannot be safeguarded in Manipur. Only for their political gain, politicians often insisted many innocent electorates to condemn the ruling government, charging them of betrayal or neglecting them in any field on communal line. Even former Chief Minister, Rishang Keishing (belonging to Tangkhul Naga), longest served chief minister ever in Manipur, claimed 'Meiteis betrayed them' (North East Sun, 2005).

On the eve of assembly elections held in 2007, the UNC, its constituent Tribe Councils/Hohos and the Naga civil societies have decided to define the objectives with which the Naga people must approach the same for securing representatives through whom the voice of the Naga people can be articulated loud, clear and consistently. The common front floated by the UNC has been christened as the United Naga Democratic Front (UNDF).[23] As usual, UNC had declared proposed instructions to be followed by Naga candidates and warned not to contest election on any political party instead independent on the eve of elections of 2007. They searched Naga aspirant candidate from those who have been loyal to them and have strong support to their cause. There was also conflict between Naga politicians sponsored by UNC and other Naga aspirant candidates.[24] At this juncture, ethnic political parties in the name of Naga unification suddenly emerged, such as Peoples Democratic Alliance (PDA), United Peoples Democratic Party (UPDP) and Naga National Party (NNP) in anticipation of getting benefit from the UNC Declaration. As there was high intensity of ethnic politics, A.B. Bardhan (a leader of the Communist Party of India) stated that 'CPI had not established units in 11 constituencies where Nagas dominate, as several candidates were fielded under the diktats of the NSCN-IM'.[25]

Subsequently, in the valley, barring the Congress party, all the parties put up Naga issue, and several changes appeared in state politics. Several parties had targeted ruling Congress party as a result of its past records and signatory issue of Congress Naga MLAs and MPs that submitted a memorandum to Prime Minister of India showing their support to Naga unification.[26]

Result and findings

Indeed, empirical analysis confirmed large number of evidences from newspapers while many of the respondents could not give the details freely, which might be the result of psycho-fear. As such the analysis seems to emphasise on hill area where ethnic assertion is very high, but it is the outcome or real picture. On the contrary, in the valley, many respondents seemed to readily respond to real picture to some extent. The state of democracy in Manipur is clear example of flawed democracy where elections are totally in servitude though high percentage of voting is witnessed. Indeed, all these malpractices are against the electoral integrity that eventually destroys democratic value. Electorates of this state cannot decide their franchise, instead they are influenced by various factors. In the backdrop of various homeland demands and ethnic conflict, government has seemingly no means to develop a concrete mechanism, instead it is handy to politicise the issue. Conflict and odds are becoming incessant and are compelling the citizens to be habitual of such pandemonium. Governance is persisting in Manipur where non-state actors have been dominating all the administration in certain ways. It is also compelled by the conflict situation. Government machineries are acting on the drum beats of the outfits. Ethnicity-based insurgency often made attempt to nominate their candidates in elections.

Concluding remark

Indeed, the real democracy is unthinkable without respecting electoral integrity within a viable political party system. In an ideal setting, political parties, political elites and forerunners should not indulge in immoral and unethical politics though it is not an easy task where politics is always associated with power game. Violation or external control of rights, whether civil, political and economic, is unacceptable from the legal or moral point of view. The matter of ethnic assertion in the valley area was very different from the hill in the sense that some ethic organisations in the hill area have been trying to put their ethnic nominee in the legislature. On the other hand, tendency of ethnic assertion was very low in the valley; the appearances on the eve of election were sometimes reflection of the Naga integration. Besides, no organisation or party stands for particular community in the valley area as Naga parties do in the hill areas. As the analysis is trying to encompass ethnic assertion, the matter of hill and valley is quite different as the appearance of valley is just reflection from hill. Free electioneering is possible only when the existing democracy permits the political activity in the 'public sphere', which mediates the realm between civil society and the state connected to both spheres, yet distinct from them and from the economy. It consists of civic associations, social movements, interest groups, the media and arenas of public opinion formation. Logically, intimidation and threat may have certain impact on the freedom of choice, but it is very unclear as the finding shows relatively different assumption and witnesses. This is the way how democratic process converts into state of plutocracy in conflict regions. Ethnic assertion is becoming core factor of various issues of social chaos in multicultural society.

Notes

1 A tiny state of India's Northeast region, home of multicultural and diverse ethnic groups, was a sovereign country and came under the British rule after 1891 Anglo-Manipuri war, later dubiously merged with Indian union in 1949. The state consists of two region: hill (90 per cent of the total area) and valley (10 per cent of the total geographical area of the state), different *tribes* (around 30 per cent of population) live in the former region and the latter region is dominated by the dominant *Meitei* community (70 per cent of the state's population).
2 A factional armed opposition group for the Naga community and under political dialogue with Government of India since 1997.
3 See, for example. the Global Commission on Elections, Democracy and Security launched by the International Institute for Electoral Assistance (International IDEA) and the Kofi Annan Foundation in March 2011.
4 Electoral Management: http://aceproject.org/ero-en/topics/electoral-management/electoral%20cycle.JPG/view accessed on 2 June 2012.
5 Interview with T. Kipgen, the founding secretary of KNA in Imphal on 13 October 2002. It became the mouthpiece of the non-Naga Hill tribes in Manipur demanding a Kuki state within the framework of Indian federation in 1960, and later on further relaxed its stand by demanding a full-fledged revenue district for the Kukis in the face of stiff Naga resistance.

6 All Kuki National Council Meeting, Resolution No. 1, held at Tujang village on 19 August 1947.
7 It was also a party that promised to work as a political platform for all the Naga people in NER. The Naga National Party (NNP) Manifesto, eighth Manipur State Legislative assembly Elections, 2002, p. 1.
8 Besides, Nagaland Chief Minister Neiphiu Rio in his election rally in Tamenglong district had stated that NPF MLAs would bring up the Naga issue in the Manipur Legislative Assembly. Staff reporter, *'O Joy cautions Rio to desist dragging integrity issue into polls'*, *Sangai Express*, Imphal, 17 January 2012.
9 NPF operated and governed in neighbouring state of Nagaland which is also popularly known for NSCN-IM's support.
10 *Imphal Free Press* under the caption of 'Choose with prudence', reported on 26 January 2012.
11 What we are experiencing during the electoral activities in the remote and hill areas, the writer interviewed and found from various officials who discharged election duty in the hill areas. One of them Baleshor, U. said that they have no option; even the security forces cannot interfere in the activities carried out by armed miscreants. Perhaps they might have to support particular candidate in such areas. Sometimes they voted as representative of all the electorates (7 July 2008).
12 It may be noted that Ng. Kumarjit Singh, a legislator, stated in the Manipur Legislative Assembly Session 1991, 'Politicians have to stop the culture of using youth by supplying armed ammunition and harassing the people for some political benefit on the eve of election. Politics should be done sincerely and genuinely.' *Assembly proceeding*, Manipur Legislative Assembly, Government of Manipur, 22 July 1991.
13 Another politician Kim Gangte said 'Voters in the hill districts failed to get their franchise . . . had been paralyzed and manipulated by militants'. Hueiyen News Service, TMC demands re-poll in hill constituencies while BJP wants re-poll in Thoubal AC, *The Hueiyenlanpao*, Imphal, 31 January 2012.
14 Ved Marwah, Former Governor, Manipur and Jharkhand delivered a lecture on Ethno-Political Situation in India's Northeast, at Centre for Strategic Analysis, Tamil Nadu, 19 May 2010.
15 Hueiyen News Service, Congress presses ECI to disqualify NPF candidates, *The Hueiyenlanpao*, Imphal, 8 February 2012.
16 Staff reporter: NSCN-IM 'poll rampage', *The Sangai Express*, Imphal, 31 January 2012.
17 Editor, Silent Elections, *Imphal Free Press*, Imphal, 23 January 2012.
18 Press Trust of India, PC admits IM hand in poll orgy, *Imphal Free Press*, Imphal, 2 February 2012.
19 Editor, Silent Elections, *Imphal Free Press*, Imphal, 23 January 2012.
20 Intolerability against armed groups was exposed, *The Imphal Free Press*, Imphal, English local daily vernacular, 20 and 21 February 2007.
21 Factional feud was also followed with different perspective. *The Manipur Mail*, Imphal, English local daily vernacular, 21 February 2007.
22 The Declaration reads, 'I wilfully support the ongoing Indo-Naga Peace talks for a negotiated and honourable solution. On being elected I will

steadfastly defend the interest of the Naga people for the integration/ unification of the Naga areas; That I will resign from the Manipur Legislative Assembly if called upon to do so by the Naga people represented by the UNC; . . . I declare that I will accept without any reservation whatsoever, the candidature of the particular person who is determined by the Naga people represented by the UNC to be the consensus candidate; . . . I shall not defect/ split/ merge to any political party without the approval of the Naga people represented by the UNC. Failing to fulfil the above Commitments and Declaration, I and my witness shall be made liable to any measures adopted by the Naga people.'

23 To them: to fulfil the inherent and democratic aspiration of the Naga people for unification of all homeland; to mobilise the peoples' fullest support of the ongoing Indo-Naga political dialogue for an honourable solution; to provide leadership to society in nation building and to pro mote the common interest of all ethnic communities in the state.

24 Some Naga aspirant electorates condemned strongly, *The Sangai Express*, a local English daily vernacular, 3 November 2006.

25 A.B. Bardhan rightly pointed out of insurgency dominance in the hill areas, *Imphal Free Press*, English local daily vernacular, 22 January 2007.

26 In May 2005, a memorandum signed by thirteen Naga parliamentarians and legislators of Manipur was submitted to the Prime Minister Manmohan Singh demanding Naga unification. Rajya Sabha MP Rishang Keishing and Lok Sabha MP Mani Charanamei along with some other ministers and MLAs of the Naga community in the then Secular Progressive Front government led by Congress party (2002–07) have closed relation with the Naga militants.

References

Bibhu, P.R. (2010). 'Manipur Assembly Elections: Divergence of Ethnic Interests, Institute for Conflict Management' (online), available at: http://www.ipcs.org/article/military/manipur-assembly-elections-divergence-of-ethnic-interests-703.html, accessed on 23 October 2010.

Eric, C.B. (2004). *Beyond Free and Fair: Monitoring Elections and Building Democracy*. Washington, DC: Woodrow Wilson Center Press.

ICCPR (1996). Adopted by the Committee at its 1510th meeting (fifty-seventh session) on 12 July 1996 (online), available at: http://www.unhchr.ch/html/menu3/b/a_ccpr.htm, accessed on 23 March 2010.

James, B. (2005). *Attention Deficit Democracy*. New York: Palgrave Macmillan, p. 23.

Karl-Heinz, N. (2009). *The Funding of Party Competition: Political Finance in 25 Democracies*. Baden-Baden: Nomos Verlag.

Kevin, C.Z. (2004). *Paying for Democracy*. Essex: ECPR Press.

Lijphart, A. (1999). *Patterns of Democracy – Government Forms and Performance in Thirty-Six Countries*. New Haven, CT: Yale University Press.

Magnus, O. and Hani, Z. (2011). *Political Finance Regulation: The Global Experience*. Washington, DC: IFES.

Marwah, V. (2010). 'Ethno-political situation in India's Northeast, CSA, Intra-state conflicts and effects', available at: http://internalconflict.csa-chennai.org/2010/05/ethno-political-situation-in-indias.html, accessed 28 September 2012.

Michael, B., Yun-han, C., and Marta, L. (2010). 'Who Votes? Implications for New Democracies', *Taiwan Journal of Democracy*, 6(1): 107–36.

Michael, K. (2011). *The Politics of Party Funding*. Oxford: Oxford University Press.

Keising, Rishang. (2005). *Meiteis Betrayed*, North East Sun, 15 July: Guwahati.

Pippa, N. (2004). *Electoral Engineering*. New York: Cambridge University Press.

——— (2012). 'Are There Global Norms and Universal Standards of Electoral Integrity and Malpractice? Comparing Public and Expert Perceptions, Perceptions of Electoral Integrity', Draft 1.0 6/1/2012 (online), available at: www.electoralintegrityproject.com, accessed on 23 February 2013.

Roy, A. K. (1986). 'KNA and the Electoral Politics in Manipur', In Dutta, P.S. (Ed.), *Electoral Politics in North East India*, Guwahati, Assam: Omsons Publications, p. 67.

Singh, L. M. (2013). 'Emerging Trends of Ethnic Assertion and Electoral Integrity in Manipur: State of Electoral Rights in Conflict Situation', Presented on 2-Day State Level Seminar on Electoral Politics in Manipur at Modern College, Imphal (Manipur) on 23 December 2013.

Index